Understanding Digital PCS
The TDMA Standard

For a complete listing of the *Artech House Mobile Communications Library*,
turn to the back of this book.

Understanding Digital PCS
The TDMA Standard

Cameron Kelly Coursey

Artech House
Boston • London

Library of Congress Cataloging-in-Publication Data
Coursey, Cameron
  Understanding digital PCS : the TDMA standard / Cameron Coursey.
    p.   cm.
  Includes bibliographical references and index.
  ISBN 0-89006-362-1 (alk. paper)
  1. Personal communications service systems.   2. Time division multiple access.
  I. Title
  TK5103.485.C68   1998
  621.3845—dc21   98-42793
                                                                                            CIP

British Library Cataloguing in Publication Data
Coursey, Cameron Kelly
  Understanding digital PCS : the TDMA standard.—(Artech House mobile communications library)
  1. Personal communication service systems—Standards
  2. Digital communications—Standards
  I. Title
  621.3'8456'0218
  ISBN 0-89006-362-1

Cover and text design by Darrell Judd

© **1999 ARTECH HOUSE, INC.**
685 Canton Street
Norwood, MA 02062

All rights reserved. Printed and bound in the United States of America. No part of this book may be reproduced or utilized in any form or by any means, electronic or mechanical, including photocopying, recording, or by any information storage and retrieval system, without permission in writing from the publisher.
    All terms mentioned in this book that are known to be trademarks or service marks have been appropriately capitalized. Artech House cannot attest to the accuracy of this information. Use of a term in this book should not be regarded as affecting the validity of any trademark or service mark.

International Standard Book Number: 0-89006-362-1
Library of Congress Catalog Card Number: 98-42793

10 9 8 7 6 5 4 3 2

Contents

Preface	xiii
1 The Digital PCS Family Tree	**1**
1.1 Advanced Mobile Phone Service	1
1.2 IS-54	6
1.3 IS-136 and Digital PCS	9
1.4 136+	13
1.5 136HS	14
1.6 Relationship of Digital PCS to GSM	15
1.7 A summary of the Digital PCS family tree	16
References	17
2 Advantages of Digital PCS	**19**
2.1 Capacity enhancement	20

2.1.1 Time Division Multiple Access	20
2.1.2 Hierarchical cell structures	24
2.1.3 Adaptive channel allocation	27
2.2 Feature flexibility	29
2.2.1 Layered protocol	29
2.2.2 Teleservices	31
2.3 AMPS compatibility	33
References	34

3 The Digital PCS Family of Standards 35

3.1 Telecommunications Industry Association	36
3.2 Interim Standards and American National Standards	38
3.3 The Digital PCS standards	39
3.3.1 IS-136	39
3.3.2 IS-137 and IS-138	41
3.3.3 IS-641	42
3.3.4 IS-130 and IS-135	43
3.3.5 TIA/EIA-136	43

4 Layer 1: The Digital PCS Physical Layer 47

4.1 Channelization and duplexing	48
4.2 Modulation format	51
4.3 Power output	54
4.4 Frame and time-slot structures	56
4.4.1 Digital control channel time-slot structure	59
4.4.2 Digital traffic channel time-slot structure	63
4.5 Channel coding and interleaving	66
4.5.1 Digital control channel	68
4.5.2 Digital traffic channel	69
References	69

5 Layer 2: The Digital PCS Data Link Layer 71

5.1 The OSI reference model	71

5.2 Layer 2 of the DCCH	73
5.2.1 DCCH layer 2 service access points and service primitives	73
5.2.2 DCCH layer 2 protocols	76
5.2.3 DCCH layer 2 media access control	82
5.2.4 ARQ mode operation	85
5.2.5 Monitoring of radio link quality on the DCCH	87
5.3 Layer 2 of the DTC	87
References	88

6 Layer 3: The Digital PCS Network Layer — 89

6.1 Introduction to layer 3	89
6.2 DCCH message set	92
6.2.1 BCCH messages	92
6.2.2 SPACH messages	95
6.2.3 RACH messages	98
6.3 DTC message set	100
6.3.1 SACCH and FACCH messages on the forward DTC	100
6.3.2 SACCH and FACCH messages on the reverse DTC	103
Reference	105

7 Network Architecture and Intersystem Operation — 107

7.1 Network overview	107
7.2 Mobile station	109
7.3 Base station	110
7.4 Mobile switching center	111
7.5 Interworking function	112
7.6 Home location register and visitor location register	113
7.7 Authentication center	114
7.8 Teleservice server	115
7.9 Network interfaces	116
References	118

8 Digital PCS Mobile Operation — 119

8.1 Mobile types 119
8.2 Mobile states 120
   8.2.1 Control Channel Scanning and Locking 120
   8.2.2 DCCH Camping 122
   8.2.3 Registration Proceeding 123
   8.2.4 Origination Proceeding 124
   8.2.5 Waiting for Order 125
   8.2.6 Terminated Point-to-Point Teleservice Proceeding 125
   8.2.7 SSD Update Proceeding 126
   8.2.8 Originated Point-to-Point Teleservice Proceeding 127
8.3 Mobile station procedures 128
   8.3.1 DCCH Scanning and Locking 128
   8.3.2 Control Channel Selection 129
   8.3.3 Control Channel Reselection 130
   8.3.4 Termination 131
   8.3.5 Origination 132
   8.3.6 Originated Point-to-Point Teleservice 132
   8.3.7 Registration-related procedures 133
   8.3.8 Authentication procedures 135
   8.3.9 Mobile Assisted Channel Allocation 136
8.4 Mobile station identities 136
   8.4.1 MIN 136
   8.4.2 IMSI 137
   8.4.3 TMSI 138
References 139

9 Reselection and Hierarchical Cell Structures 141

9.1 Neighbor cells and neighbor cell information 142
9.2 Control Channel Locking and Reselection Criteria procedures 143
9.3 Reselection Trigger Conditions 144
9.4 Candidate Eligibility Filtering 146
9.5 Candidate Reselection Rules 155
9.6 Reselection examples 157

9.6.1 Radio Link Failure	158
9.6.2 Cell Barred	159
9.6.3 Server Degradation	161
9.6.4 Directed Retry	162
9.6.5 Priority System	163
9.6.6 Periodic Evaluation	165
9.7 Implementing hierarchical cell structures	165
Reference	169

10 Mobile Sleep Mode 171

10.1 Paging frame class and PCH subchannels	172
10.2 Processes impacting sleep mode	175
10.2.1 Neighbor channel measurements	175
10.2.2 Mobile Assisted Channel Allocation	178
10.2.3 Rereading the broadcast control channel	179
10.3 Extending sleep mode with optional enhancements	181
References	182

11 Voice Services 183

11.1 Voice coding and decoding	183
11.1.1 IS-641 ACELP	186
11.1.2 Discontinuous transmission and comfort noise	191
11.1.3 Vocoder implementation	193
11.2 Call processing	194
11.2.1 Call establishment	194
11.2.2 Call handoff	199
11.2.3 Vocoder assignment	201
References	203

12 Teleservices 205

12.1 Introduction	205
12.2 Teleservice delivery	207
12.3 SMS_TeleserviceIdentifier and Higher Layer Protocol Identifier	211

12.4	Teleservice segmentation and reassembly	213
12.5	Broadcast teleservice transport	213
12.6	Defined teleservices	217
	12.6.1 Cellular messaging teleservice	218
	12.6.2 Over-the-air activation teleservice	223
	12.6.3 Over-the-air programming teleservice	227
	12.6.4 Generic UDP transport teleservice	229
	References	230

13 Circuit-Switched Data Services 231

13.1	Introduction	231
13.2	Analog circuit-switched data service	231
13.3	Digital circuit-switched data service	233
	13.3.1 Radio Link Protocol 1	235
	13.3.2 IS-135 data part	238
	13.3.3 Data call processing	240
	References	241

14 Nonpublic Services 243

14.1	Network types	243
14.2	Private and residential system identities	245
14.3	Autonomous systems	249
	References	253

15 Special Considerations for 1,900-MHz Operation 255

15.1	Intelligent roaming	255
15.2	Hyperband reselection and handoff	267
15.3	Dual-band, dual-mode mobiles	270

16 Authentication, Privacy, and Encryption 273

16.1	CAVE	273
16.2	A-Key and SSD	274
16.3	Authentication procedures	275

16.4	Updating shared secret data	280
16.5	Voice privacy and signaling message encryption	282
16.6	Encryption on the DCCH	284
	References	285

17 Network Parameter Settings 287

17.1 Importance of network parameter settings 287
17.2 Selection and access parameters 291
    17.2.1 DCCH location 291
    17.2.2 Access thresholds 296
    17.2.3 Access attempts 299
    17.2.4 DCCH structure 302
17.3 Reselection and handoff boundaries 304
References 305

18 Equipment Testing 307

18.1 Why test? 307
18.2 Network testing 308
18.3 Mobile testing 309
    18.3.1 Performance testing 311
    18.3.2 Interoperability testing 314
    18.3.3 Protocol conformance testing 316
    18.3.4 CTIA certification program 322

19 Towards Global TDMA 325

19.1 The Universal Wireless Communications Consortium 325
19.2 Global TDMA Forum 326
19.3 Global WIN Forum 328
19.4 Global Operators' Forum 329

20 The Future of Digital PCS 331

20.1 IMT-2000 and third-generation cellular 331
20.2 136+ 333

20.3 136HS	335
20.4 Potential further Digital PCS enhancements	336
References	337

Appendix A: UWCC Member Companies — 339
A.1   Board-Level Companies — 339
A.2   General Membership — 340

Glossary — 343

About the Author — 367

Index — 369

Preface

DIGITAL PERSONAL COMMUNICATIONS service (PCS) is a popular marketing name for the most successful Time Division Multiple Access (TDMA) cellular standard originating in North America. Some of the largest cellular service providers in the world, including AT&T Wireless Services, BellSouth Cellular Corporation, Rogers Cantel, and SBC Communications, have deployed Digital PCS systems on a large scale. Network suppliers such as Ericsson, Hughes Network Systems, Lucent Technologies, and Nortel have developed Digital PCS systems. The largest mobile phone suppliers, including Ericsson, Motorola, NEC, Nokia, and Philips have developed mobiles to support Digital PCS. By the end of 1997, it was estimated that there were over 15 million Digital PCS subscribers worldwide.

The actual Digital PCS standard is known as TIA/EIA-136 and was developed by the Telecommunications Industry Association (TIA). TIA/EIA-136 grew out of the first digital cellular standard for North America, IS-54 (IS for Interim Standard). The phenomenal success of analog cellular, known as Advanced Mobile Phone Service (AMPS), led directly to the rise of IS-54, and later TIA/EIA-136. By the late 1980s, capacity shortfalls were experienced in many AMPS markets. A method for increasing cellular capacity was needed, and IS-54 offered a threefold increase in capacity over AMPS.

While IS-54 offered the cellular service provider greater network capacity, it did not offer the customer any significant advantages over AMPS, and it cost more. There was no real incentive for customers to switch to digital. TIA/EIA-136 was developed to change that. With TIA/EIA-136, the cellular service provider still realizes at least a threefold increase in capacity, while the customer can take advantage of advanced features such as privacy, longer battery life, caller ID, message waiting indication, and short message service. When the Federal Communications Commission (FCC) allocated a new block of frequencies in the 1,900-MHz range for PCS, the TIA/EIA-136 standard was expanded to support operation at these frequencies. TIA/EIA-136 systems are now operational in both the 800-MHz and 1,900-MHz bands.

This book explains the TIA/EIA-136 standard in-depth, and describes how TIA/EIA-136-compliant mobile phones and cellular networks interoperate. It is intended for a wide-range of readers, from the design engineer who desires to understand how the piece or part he or she is responsible for fits into the whole system, to the RF performance engineer in the field setting parameters, to the sales agent explaining the advantages of Digital PCS to a prospective customer.

The chapters are written and organized to allow the reader to focus on those topics of greatest interest. Chapter 1 summarizes the history of cellular technology and provides a backdrop for the remainder of the text. Chapter 2 describes the advantages of TIA/EIA-136, including capacity enhancement, feature flexibility, and compatibility with AMPS. Chapter 3 details the standardization process for TIA/EIA-136. In Chapters 4, 5, and 6, the layered architecture of TIA/EIA-136 is described. Chapter 7 presents a description of the overall cellular network, including the switching and interworking functions. A description of how a TIA/EIA-136 mobile phone operates is provided in Chapter 8. Chapter 9 describes hierarchical cell structures, which allow the cellular service provider to layer cells and increase network capacity even more. The concept of sleep mode, which increases mobile phone battery life, is presented in Chapter 10. Chapter 11 describes how the mobile phone and the network work together to allow customers to place and receive calls, and describes the voice coding technology used with TIA/EIA-136. Teleservices, a powerful set of TIA/EIA-136 features that speeds the introduction of new services to the customer, are explained in Chapter 12. The data capabilities of TIA/EIA-136 are examined in Chapter 13. Chapter 14 describes what is known as TIA/EIA-136 nonpublic service, which can be used to provide unique features to business and residential customers. The modifications made to TIA/EIA-136 to support operation in the 1,900-MHz PCS bands are explained in Chapter 15. Security aspects of TIA/EIA-136 are described in Chapter 16. Chapter 17 describes the network parameter settings used to optimize Digital PCS system performance, while Chapter 18 focuses on the Digital PCS network and mobile phone testing. The industry activities geared to make TIA/EIA-136 a global system are presented in Chapter 19. Finally, Chapter 20 describes planned and potential enhancements to TIA/EIA-136 to take Digital PCS into the twenty-first century.

I would like to acknowledge the developers of Digital PCS for their hard work and technical competence—from the technologists and standards representatives who formulated the TIA/EIA-136 standard, to the designers in the supplier community who continue to build products to the standard, to the operational staff in the service provider community who make it all work. I would like to thank SBC for supporting the writing of this book. In particular, thanks go to Brad Bridges and Greg Williams at SBC Technology Resources for giving me the opportunity to participate in the development of Digital PCS. Thank you to my family, Karen, Nathanael, and Joshua, for putting up with the long hours that I spent with this book instead of with you. Finally, to Him who is able to do immeasurably more than I can ask or imagine, according to His power at work within me, to Him be glory in Christ Jesus forever and ever.

The Digital PCS Family Tree

1.1 Advanced Mobile Phone Service

It is appropriate to begin a description of Digital PCS with its grandfather, the Advanced Mobile Phone Service (AMPS). AMPS was the first cellular system deployed in the United States, and has been deployed throughout the world. AMPS users still make up a large percentage of the cellular subscriber base. AMPS is a cellular radio telephone system that uses Frequency Division Multiple Access (FDMA) in a cell site and frequency reuse among cell sites to maximize the number of simultaneous users in a given spectrum allocation. With FDMA, users desiring service simultaneously in a cell are assigned channels on different frequencies. With frequency reuse, a channel is used multiple times throughout a service area, and the reuse distance is governed by the required carrier-to-interference ratio (C/I) to maintain acceptable quality. AMPS is often referred to within the technical community by the name of the standard developed by the Telecommunications Industry Association (TIA) and Electronics

Industry Association (EIA) to describe its air interface, EIA/TIA-553. EIA/TIA-553 specifies the use of frequency division duplex (FDD) operation with 30-kHz RF channels in both the forward (from the base station to the mobile) and the reverse (from the mobile to the base station) directions, with a 45-MHz separation between the forward and reverse channels. EIA/TIA-553 also specifies the protocol used by the mobile and the base station to communicate with one another. The companion American National Standard TIA/EIA-41 specifies the network interface between AMPS systems to allow cellular subscribers to roam between cellular networks and still obtain service.

In 1983, the Federal Communications Commission (FCC) in the United States allocated 40 MHz of spectrum in the 800-MHz band for cellular service. The spectrum was divided equally between two providers in each service area. Later, the FCC allocated an additional 10 MHz—5 MHz for each service provider—for cellular use. AMPS was the cellular technology deployed in this spectrum. AMPS was also adopted as the first-generation cellular technology for Canada and Mexico, and for the majority of South America. AMPS has also been widely deployed in Asia and Australia.

From the mobile perspective, AMPS works in the following way [1]. EIA/TIA-553 specifies the use of analog control channels (ACCs) to provide signaling and control. ACCs use frequency shift keying (FSK) modulation and send data at a rate of 10 Kbps. Each cell site broadcasts an ACC on 1 of 21 frequencies. The ACC contains overhead messages that provide information about the system. When an AMPS mobile powers up, it scans the 21 ACCs in the home band, selects the one with the strongest received signal strength, and reads the overhead message. If the mobile cannot synchronize to the ACC with the strongest received signal strength on the home band, it attempts to synchronize to the ACC with the second strongest signal. If it still cannot synchronize, it scans the 21 ACCs in the nonhome band and goes through the same process. Once the mobile has determined the appropriate system to obtain service from, it sets its internal parameters according to what it has read on the overhead messages and enters what is called the idle task on the ACC. At this time, the mobile typically seizes the reverse ACC and registers with the system. The mobile then begins monitoring the ACC for mobile station control messages directed to its mobile identification number (MIN). If the mobile moves out of range of the current ACC, it rescans all 21 ACCs on the current band, selects the one with the strongest received signal strength, and begins to monitor it for messages. The mobile is typically not required to reregister when it changes ACCs in the same service area, but is often required to reregister periodically so the network knows that it is still in service. This process is flowcharted in Figure 1.1.

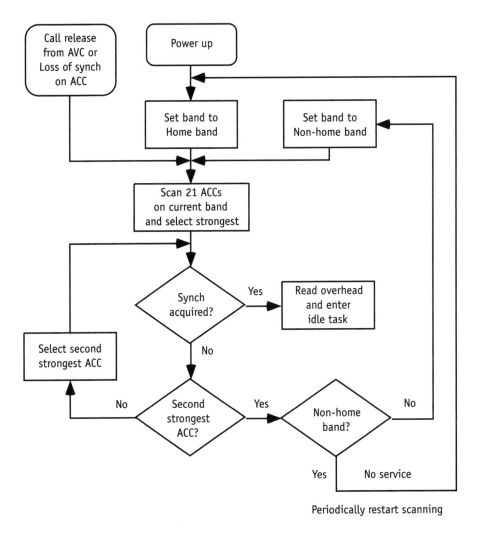

Figure 1.1 How an AMPS mobile finds an ACC.

If the mobile receives a page message while monitoring the ACC, it rescans all 21 ACCs, selects the strongest ACC, responds to the page on this ACC, and awaits an analog voice channel (AVC) assignment. EIA/TIA-553 specifies the use of frequency modulation (FM) on the AVC for voice services. When an AVC assignment message is received on the ACC, the mobile tunes to the appropriate channel and begins transmitting signaling tone on the reverse AVC. The base station then sends an alert on the forward AVC, and the mobile alerts the user of the incoming call. If the user answers the call, alerting stops, the mobile stops sending signaling tone, the end-to-end voice connection is completed, and the conversation begins. A similar procedure is used for mobile-originated calls except that the mobile sends an origination message on the reverse ACC (after conducting a rescan of the 21 ACCs and selecting the one with the strongest received signal strength) and then awaits and AVC assignment. No alerting is provided to the mobile user originating the call, of course. If the mobile moves outside of the coverage area of one cell site while a conversation is occurring on the AVC, the base station sends a handoff message to the mobile instructing it to change channels. The mobile acknowledges the order on the current AVC and tunes to the new AVC that is transmitted from a different base station. When the conversation ends, the mobile once again scans all 21 ACCs and selects the most appropriate one to monitor, as described in the previous paragraph.

From the network perspective, AMPS works in the following way (refer to Figure 1.2 for a diagram of the AMPS network). In its simplest form, the network is composed of base stations (BSs) located at cell sites, a mobile switching center (MSC), a home location register (HLR), and a visitor location register (VLR). The MSC has signaling links into the BSs, the HLR, and the VLR, and voice trunks into the BSs and the public switched telephone network (PSTN). Different cellular systems communicate with each other over signaling links using the TIA/EIA-41 protocol, described in Chapter 7. The example used here will assume the mobile is operating in its home system. When a mobile powers up and sends a registration message to a BS, the BS relays the registration to the MSC over a signaling link. The MSC confirms with the HLR that the mobile user has an active subscription, the location of the mobile is updated in the HLR, a VLR record is created for the mobile, and the mobile is considered registered. When an incoming call to the

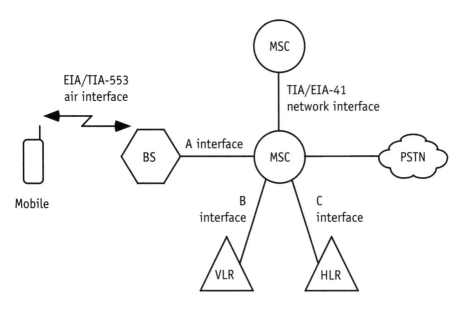

Figure 1.2 The AMPS network.

subscriber's directory number is routed from the PSTN to the MSC, the MSC queries the VLR for the mobile's registration status and the subscriber's features. The VLR returns information to the MSC indicating the mobile is located in the MSC's serving area. The MSC directs the BSs to page the mobile. The BSs formulate the page message and transmit it as a mobile station control message on the ACC. When the mobile responds to the page, the BS that receives the response notifies the MSC of this fact. At this point, the voice circuit is established from the MSC to the BS, the BS begins transmitting an AVC, and the BS assigns the mobile to the AVC. The call is connected when the BS detects a signaling tone from the mobile on the AVC. When a mobile sends an origination message to a BS on an ACC, the BS signals the MSC of the origination, establishes a voice circuit with the MSC, and assigns the mobile to an AVC. The MSC confirms the mobile's service qualification and features, then routes the call to the PSTN for delivery to the called party.

The specific AMPS handoff algorithms and procedures used by the network are not standardized. Typically, the BS monitors the received signal strength of the mobile. When the received signal strength falls below a threshold, the MSC instructs neighboring base stations to tune to the assigned AVC and measure the received signal strength from the mobile. The MSC then determines whether the mobile would be best served by remaining connected to the current base station or handing off to a neighboring base station. If the MSC determines that a handoff is appropriate, the target BS is instructed to accept the handoff. The target base station establishes a voice circuit with the MSC, begins transmitting an AVC, and informs the MSC of the AVC channel. The MSC instructs the serving BS to initiate a handoff of the mobile to the new channel. The BS issues a handoff order to the mobile and informs the MSC when the mobile acknowledges the handoff with signaling tone. At this point, the voice circuit between the MSC and the serving BS is torn down and the target BS becomes the serving BS.

1.2 IS-54

If AMPS is the grandfather of Digital PCS, IS-54 is its father. IS-54 became the American National Standard TIA/EIA-627, but is still better known as IS-54. It has gone by other names as well, such as North American Digital Cellular (NADC), U.S. Digital Cellular (USDC), and Digital AMPS (D-AMPS). IS-54 was developed in 1991 to address a problem brought on by the success of AMPS—a capacity shortfall in the largest AMPS markets. As the number of AMPS subscribers increased, there became a shortfall of radio resources in high-traffic areas. Call blocking rates increased in these high-traffic areas, particularly during the busy hours in the morning and afternoon as subscribers made their way to and from work. IS-54 offered three times the capacity of AMPS without requiring major network changes, which was attractive to some of the largest cellular service providers.

As shown in Figure 1.3, IS-54 adds one more type of channel, called the digital traffic channel (DTC), to AMPS [2]. A single 30-kHz RF channel contains three full-rate DTCs. A user is assigned one of the full-rate DTCs for voice service, so a single RF channel can support three users simultaneously. This is accomplished by splitting the 30-kHz RF channel into time slots. With a full-rate DTC, a user has access to the 30-kHz RF channel for approximately 6.7 ms every 20 ms. This type of operation is

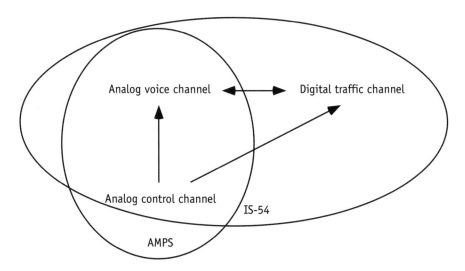

Figure 1.3 A comparison of IS-54 and AMPS channels.

called Time Division Multiple Access (TDMA). TDMA requires the use of a digital modulation format. The digital modulation scheme used in IS-54 is $\pi/4$ shifted differentially encoded quadrature phase shift keying ($\pi/4$ DQPSK). Voice coding must be used to compress the user's voice for transmission over a DTC because the user no longer has access to the entire bandwidth of the channel. The voice coding scheme used in IS-54 is known as vector sum excited linear predictive (VSELP) coding. VSELP is a form of voice coding known as source coding, in which the speech is modeled at the encoder and synthesized at the decoder. IS-54 is a dual-mode standard, and all IS-54 mobiles are dual-mode mobiles. This means that an IS-54 mobile can operate on both an AMPS network using an AVC and an IS-54 network using a DTC.

IS-54 works similar to AMPS, with the only major differences being in the assignment of a channel for voice services and the handoff between voice channels. When a DTC is assigned to a mobile by a base station, both the channel number and the time slot of the DTC must be provided. This same information must be provided in the handoff message. The handoff procedure takes advantage of feedback from the mobile in the form of channel quality measurements on the current channel and neighboring channels in a process called mobile assisted handoff (MAHO).

When a mobile is assigned to a DTC, the base station sends the mobile an order to start measurements on the current channel and a list of channels that are provided. The mobile measures the received signal strength and the bit error rate (BER) on the current channel, and the received signal strength on the remaining channels. It can measure signal strength on these other channels during the period when it is not transmitting or receiving on the DTC. The mobile reports these measurements to the base station, and the MSC uses these measurements to assist in the handoff process. In this way, the MSC has information on both the reverse and forward links when it makes handoff decisions.

Another major enhancement of IS-54 over AMPS is the addition of *authentication, signaling message encryption*, and *voice privacy*. While it is optional for the network and service provider to use these fraud prevention features, they are required to be supported by every IS-54 mobile. Authentication is the process of validating that a mobile is authentic and not a clone. With signaling message encryption, sensitive subscriber information on the AVC and DTC may be protected with a cipher. Voice privacy, which is only available on the DTC, masks the conversation from eavesdropping. Authentication, signaling message encryption, and voice privacy are all based on the use of a secret key, called an *A-key*, that is known only to the mobile and the home system. Because these fraud prevention features are also used in Digital PCS, they will be explained in detail in Chapter 16.

One final key enhancement over AMPS is provided by IS-54: calling number identification. This feature has proven to be one of the most popular for cellular subscribers because it lets them know the telephone number of the person calling them on their mobile, an important piece of information as the subscriber typically must pay for receiving calls in addition to originating them. Calling number identification is provided to an IS-54 mobile by the base station once the mobile is assigned to an AVC or a DTC. The calling number typically is displayed to the mobile user at the same time the user is alerted of an incoming call. If the user chooses not to accept the call, the mobile turns off its transmitter and returns to the ACC. The MSC may then perform secondary call treatment on the incoming call, such as delivering the call to voice mail.

Even with the new features enabled by IS-54, and the promise of increased capacity for the service provider, IS-54 did not achieve the anticipated success. The key factor that contributed to the failure of IS-54 to achieve high penetration in most markets was IS-54 voice quality. The IS-54 voice coder, VSELP, provides adequate voice quality under clean speech and clean channel conditions. Clean speech and clean channel conditions are rarely encountered in the cellular environment, however. Speech is often mixed with a high level of background noise in the mobile environment. Voice coders (vocoders) are designed to model speech, not other signals. In high-background-noise conditions, VSELP exhibits a phenomenon called "swirling" or "the waterfall effect," so named because of the sound that is produced as a result of the noise. This makes the voice sound artificial. Preprocessing of the speech to minimize the background noise can reduce the effects of swirling, but this was not implemented in the early days of IS-54. Under degraded channel conditions—for example, interference, weak signal, or high multipath fading—the BER of the coded voice signal can be high enough that a frame error is declared and an entire frame (20 ms) of speech is lost. When the frame error rate (FER) is high enough, muting of large segments of speech occurs and the conversation is unintelligible. While the service provider can design the network to minimize the FER, it often requires the addition of new cell sites and a looser frequency reuse, which mitigates some of the capacity advantages of IS-54. This latter effect is a problem for existing service providers with networks already designed to meet the C/I required for AMPS.

1.3 IS-136 and Digital PCS

Starting with IS-136 Revision 0 published in 1994 [3], the IS-136 (now American National Standard TIA/EIA-136) family of standards brings a wealth of new features, solves the voice quality issue with IS-54, and provides support for operation in the 1,900-MHz frequency band allocated around the world for personal communication service (PCS). IS-136 Revision 0 standardized the digital control channel (DCCH), the key to most of the new features. IS-136 Revision A (IS-136-A) standardized a more robust and higher fidelity voice coder (vocoder), as well as operation in the 1,900-MHz *hyperband*. Digital PCS began with IS-136 Revision 0. The majority of Digital PCS networks have been upgraded to IS-136-A (the same as TIA/EIA-136), however.

Table 1.1 compares the features found in AMPS and IS-54 with those of Digital PCS. High-quality digital voice is achieved through the use of a new vocoder based on algebraic code excited linear predictive coding (ACELP). This ACELP vocoder has been standardized as IS-641. IS-641 operates at a voice coding rate of 7.4 Kbps, compared to a voice coding rate of approximately 8 Kbps for VSELP. Even though the voice coding rate is lower, IS-641 consistently scores higher in subjective voice quality testing than VSELP under clean and background noise conditions. The lower voice coding rate of IS-641 makes it possible to add more error detection and correction coding to every frame of transmitted speech. This makes IS-641 more robust to channel impairments than VSELP. The IS-641 voice coding process is described in Chapter 11.

The majority of the remaining new features listed in Table 1.1 for Digital PCS are made possible by the DCCH. The DCCH performs the same type of signaling and call control functions as the ACC, but uses the digital modulation and frame structure of the DTC. It therefore does not

Table 1.1
AMPS, IS-54, and Digital PCS Features

AMPS	IS-54	Digital PCS
Analog voice service	AMPS features plus:	IS-54 features plus:
Analog circuit-switched data service	Digital voice service with VSELP	High-quality digital voice service with IS-641 ACELP
	Voice privacy	Sleep mode for increased standby time
	Authentication and signaling message encryption	Message waiting indication
	Calling number identification	Short message service
		Nonpublic service
		Dual-band operation
		Over-the-air activation
		Digital circuit-switched data service
		Calling name identification

occupy a full RF channel, but only a full-rate digital channel (two time slots out of six). The DCCH is composed of a number of logical channels in the forward direction and one logical channel in the reverse direction. The logical channels in the forward direction are divided into broadcast channels and point-to-point channels, while the reverse channel is a point-to-point channel. The broadcast channels contain information on the structure of the DCCH, the system identity, registration parameters, and other information similar to the overhead messages on the ACC.

Two key differences between the ACC and the DCCH for the mobile are the existence of a *neighbor list* on the DCCH and the concept of *paging frame class* (PFC) used on the DCCH. First, every DCCH contains a list of neighboring control channels. These neighboring control channels can be ACCs, DCCHs, or both. The mobile periodically measures the received signal strength on the neighboring control channels and evaluates whether another control channel would be the better server at any given time. If a better control channel, either DCCH or ACC, is found by the mobile upon periodic evaluation, the mobile tunes to the new control channel, synchronizes, and begins to monitor it for pages and other messages. This process of selecting the best DCCH based on the evaluation of neighbors is called *reselection*. With reselection, the mobile is guaranteed to always be on the best control channel from which to obtain service. Thus, when a mobile is operating on a DCCH, there is no need for it to conduct a rescan of the control channels prior to originating or receiving a call, as is the case for AMPS and IS-54. This shortens the call setup time. Reselection is also a more graceful way for the mobile to find a new control channel, and its use minimizes the probability that the mobile will miss a page on the control channel while it is rescanning. The reselection process is covered in Chapter 9.

The use of PFCs on the DCCH makes it possible for Digital PCS mobiles to conserve battery life by periodically entering sleep mode while operating on the DCCH. The PFC identifies how often the mobile must wake up to look for a page. The mobile looks for a page on a time slot called its assigned paging channel. At the lowest PFC, PFC1, the assigned paging channel occurs once every 1.28s. This means the mobile must only tune its receiver to the DCCH (typically) for approximately 6.67 ms, plus any additional time for synchronization, every 1.28s. When the mobile is not monitoring its assigned paging channel, it performs other tasks such as monitoring and evaluating neighbor list entries. But there is still time for the mobile to shut down its receiver and as much logic as possible to minimize current drain for some period of time. Digital PCS mobiles can obtain more than twice the standby time of AMPS or IS-54 mobiles while operating at PFC1. In the future, higher PFCs can be used to lengthen the interval between paging channel occurrences, thereby allowing the Digital PCS mobile even more time to sleep and conserve its battery. Chapter 10 describes more fully how sleep mode works in Digital PCS.

Short message service (SMS) and over-the-air activation (OAA) are examples of features that use *teleservices* on the DCCH. Teleservices are applications that use the air interface and network interface as the bearers for their transport between a network element called a teleservice server and a mobile. The remainder of the network simply acts as a relay for the teleservice, hence the name relay data, or R-DATA, for the message that is used on the DCCH to transport teleservices. For the case of SMS, the teleservice server is called a message center, and the teleservice is called cellular messaging teleservice (CMT). For the case of OAA, the teleservice server is called an over-the-air activation function (OTAF), and the teleservice is called over-the-air activation teleservice (OATS). With SMS, a mobile subscriber can be sent a short alphanumeric message over the DCCH (or DTC), and the message appears on the mobile's screen much like an alphanumeric pager. With OAA, a new subscriber's mobile can be programmed with a MIN and home system information remotely. This opens up new distribution channels for Digital PCS service. Teleservices are a powerful aspect of Digital PCS, and they are covered in greater detail in Chapter 12.

A unique feature of Digital PCS is the ability to support different network types, including public, private, and residential networks. Public networks are the traditional type of cellular networks, in which every mobile authorized for service is allowed to access every cell site. Private and residential networks, on the other hand, offer the capability to define closed user groups and wireless virtual private networks. For example, a private network could be established covering a building or campus and only accessible by business users, or accessible to these users at a discounted rate compared with public users. Every Digital PCS mobile supports this type of operation, known as *nonpublic service*. Chapter 14 describes how nonpublic service works in Digital PCS.

The remainder of the features listed for Digital PCS in Table 1.1 are also described herein. The message waiting indication (MWI) is used to notify a mobile subscriber of the type and number of messages waiting. The message type can be voice message, fax message, or SMS message. Dual-band operation is supported in Digital PCS by defining reselection and handoff capability between the 800-MHz and 1,900-MHz hyperbands. Chapter 15 describes dual-band operation in Digital PCS. Circuit-switched digital data is supported in full-rate, double full-rate, and triple full-rate modes at usable data rates from 9.6 Kbps for full-rate to 28.8 Kbps for triple full-rate. Chapter 13 describes the standards that apply for circuit-switched digital data. Finally, calling name identification is provided to the mobile over the air interface in a manner similar to calling number identification in IS-54.

1.4 136+

136+ is the next progression of Digital PCS. Why is another revision to TIA/EIA-136 necessary? First, 136+ brings additional features and functionality to cellular subscribers that were not possible previously. New functionality is often enabled as the processing and storage capabilities of mobiles and infrastructure increase. Second, 136+ is necessary to keep pace with other cellular technologies that have projects similar to 136+. Third, 136+ fulfills many of the requirements identified by the International Telecommunications Union (ITU) for third-generation cellular systems. Recognition by the ITU as a viable third-generation cellular system further opens the global market for 136+.

New features with 136+ include higher quality voice service, packet data service, a standard method of intelligent roaming, teleservice segmentation and reassembly, support for Web browsers, and broadcast SMS. Higher quality voice service is provided in two ways. A robust modulation format is used with the IS-641 vocoder to enhance the voice quality under interference conditions, and a high-level modulation format is used with a higher bit rate vocoder to provide greater fidelity in more benign RF environments. Packet data service is provided over both new modulation schemes and is tightly integrated with the voice network. The high-level modulation allows a single user data rate of up to 43.2 Kbps for packet data service. Intelligent roaming is the process used by a mobile to automatically obtain service on the best possible service provider in a given area, and is indispensable for dual-band mobiles. An interim version of intelligent roaming was used with IS-136 Revision A mobiles. The standard version of intelligent roaming for 136+ allows mobiles to obtain the best possible service faster than in the interim version, and gives the home service provider more flexibility to identify and direct the mobile to preferred service providers (see Chapter 15 for a detailed description of intelligent roaming). Teleservice segmentation and reassembly allows larger messages to be built up from smaller ones. Support for Web browsers is provided by a new teleservice called the general UDP transport service (GUTS). Finally, broadcast SMS provides the capability to send the same message to all mobiles, or a group of mobiles, in the same cell site at the same time. 136+ is undoubtedly not the last enhancement to the 30-kHz channel.

1.5 136HS

136HS is the high-speed data component of Digital PCS. Its goal is to fulfill the high-speed data requirement of a third-generation cellular system as mandated by the ITU. The ITU requirements are the support of peak user data rates of 144 Kbps for the high-speed vehicular environment, 384 Kbps for the low-speed vehicular and pedestrian environments, and 2 Mbps for the indoor office environment. These rates are only possible via the use of much higher bandwidths than the 30 kHz that traditional Digital PCS uses. In addition to these peak user data rate requirements, the industry team tasked with developing 136HS placed additional limitations on the high-speed data air interface, including backward compatibility with the TIA/EIA-136 family of standards, the capability to deploy 136HS as an overlay to existing Digital PCS systems, and the use of no more than 1 MHz of spectrum to deploy the high-speed data network in the vehicular and pedestrian environments.

Two air interfaces were selected for 136HS: one for the vehicular and pedestrian environment and another for the indoor environment. For the vehicular and pedestrian environments, a 200-kHz RF channel is split into eight time slots and a high-level modulation is used to achieve the required peak user data rates. The outdoor component of 136HS achieves a frequency reuse of 1/3, so the system can be deployed in 600 kHz of spectrum. For the indoor environment, a single wideband TDMA channel of 1.6-MHz bandwidth is used with a high-level modulation to achieve the required peak user data rates. 136HS is covered in greater detail in Chapter 20.

1.6 Relationship of Digital PCS to GSM

Other TDMA standards for cellular radio telephone systems have been successful around the world, most notably the Global System for Mobile Communications (GSM) and the upbanded versions of GSM known as DCS1800 in Europe and PCS1900 in the United States. GSM might be considered a cousin to Digital PCS. At the physical layer, the main similarity is in the TDMA structure of the channel. In fact, 136HS uses the same frame structure as GSM (and even the same modulation as a fallback mode). Another area of similarity is in the area of voice services. The IS-641 vocoder used in Digital PCS is a variant of the vocoder used in PCS1900, called US1. US1 is also used to provide the high-fidelity voice service for 136+. Still another similarity is in radio resource management [4]. Both GSM and Digital PCS use the concept of a control channel providing neighbor list information that the mobile uses to identify the best server at any given time. Both systems also have the mobile monitor a paging subchannel of a control channel for pages directed to it, thus enabling the mobile to enter sleep mode when not monitoring its assigned paging subchannel or performing other tasks. The call establishment procedures between the mobile and the base station are similar in GSM and Digital PCS, and both systems use MAHO for handoff (called handover in GSM). Many of the services supported by GSM are supported in a similar manner by Digital PCS. Teleservices such as short message service are supported in both, for example. The network architecture and mobility management of GSM and Digital PCS are even similar, although the concept of a base station controller is not present in the Digital PCS standards.

The major differences between GSM and Digital PCS, aside from the physical layer, fall into the categories of subscriber identity and the interfaces and protocols used between network elements. In GSM, a subscriber identity module (SIM) is used to store subscriber-specific information, such as the *international mobile station identity* (IMSI) and the authentication key [5]. The IMSI is comparable to the MIN used in Digital PCS, and Digital PCS allows for the use of IMSIs. GSM does not allow for the use of MINs, however. Also, Digital PCS does not have the concept of a SIM. In Digital PCS, all subscriber and home system information is stored in the *numeric assignment module* (NAM) of a mobile. The NAM is housed in nonvolatile memory within the mobile, and a mobile may contain multiple NAMs to take on different identities. This is useful if the mobile user often roams between two cellular systems and desires to have a local MIN for both systems. While the authentication process is similar between GSM and Digital PCS, the authentication keys and algorithms are different. With respect to the network interfaces and protocols, the most important difference is in the intersystem signaling. Digital PCS uses the TIA/EIA-41 mobile application part (MAP), while GSM uses the GSM MAP. If a multimode mobile were to be built that supported both Digital PCS and GSM, an *interworking function* would be required to translate between the TIA/EIA-41 and GSM MAPs for roaming to occur between the different networks [6]. Such interworking functions between GSM and TIA/EIA-41 have been developed.

1.7 A summary of the Digital PCS family tree

Figure 1.4 summarizes the relationship between Digital PCS and other cellular radio telephone systems. There is a clear path from AMPS through IS-54 to the TIA/EIA-136 family of standards. Every new generation builds upon the earlier generation and maintains backward compatibility; for example, a Digital PCS mobile can operate on an AMPS network. The compatibility between Digital PCS and AMPS will be explored more fully in the next chapter.

The Digital PCS Family Tree   17

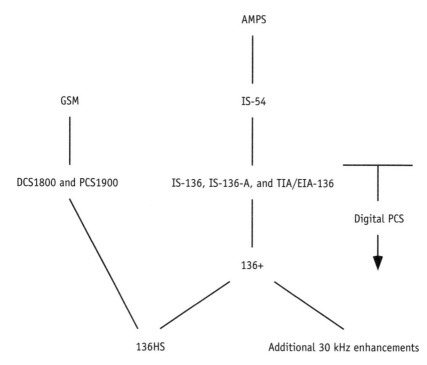

Figure 1.4   The genealogy of Digital PCS.

References

[1] American National Standards Institute, EIA/TIA-553, *Mobile Station—Land Station Compatibility Specification*, Sept. 1989.

[2] Telecommunications Industry Association, TIA/EIA IS-54-B, *Cellular System Dual-Mode Mobile Station—Base Station Compatibility Standard*, Apr. 1992.

[3] Telecommunications Industry Association, TIA/EIA IS-136, *800 MHz TDMA Cellular-Radio Interface—Mobile Station—Base Station Compatibility Standard*, Dec. 1994.

[4] Mouly, M., and M. Pautet, *The GSM System for Mobile Communications*, Cell and Sys., 1992, pp. 309–430.

[5] Ibid., pp. 67–71.

[6] Gallagher, M. D., and R. A. Snyder, *Mobile Telecommunications Networking with IS-41*, New York: McGraw-Hill, 1997, pp. 347–356.

Advantages of Digital PCS

WHY WOULD A service provider with a perfectly good AMPS network invest in Digital PCS technology? Why would a new service provider starting from scratch in virgin spectrum choose Digital PCS? Why would an existing cellular user trade in his current mobile for a new Digital PCS mobile? What would make a new cellular user select Digital PCS service over AMPS? The answer is the same for all these questions: Digital PCS provides advantages for both the service provider and the user, among them being capacity enhancement, feature flexibility, and compatibility with AMPS. At first glance, capacity enhancement might be thought of as an advantage only for the service provider. However, if the capacity enhancement equates to a higher probability that the user's call will go through, it is an advantage to him or her also. Similarly, compatibility with AMPS might appear to only benefit the service provider because it eases the conversion to digital technology. However, it is also beneficial for the user to have a single mobile that can be used in different environments, particularly for a user who roams. Conversely, feature flexibility might seem to be an advantage only to the user who can take advantage of the new features. Service providers also benefit from feature flexibility, however, because it increases their service offerings and attracts new customers. Capacity enhancement is made possible by the TDMA nature of Digital PCS, the use of hierarchical cell structures (HCS), and the use of adaptive channel allocation (ACA). Feature flexibility is enabled by the layered protocol of the Digital PCS standard and the use of teleservices. Compatibility with AMPS is achieved through the ability of the technologies to coexist with one another.

2.1 Capacity enhancement

2.1.1 Time Division Multiple Access

TDMA is an effective method of allowing more than one use of a physical resource at the same time for different purposes. An example of the use of TDMA is a political debate in which two politicians are given equal time to respond to questions. The politicians take turns answering the questions, and a timer controls the length of time that one politician is allowed to speak before relinquishing the floor to the other. Each politician knows when it is his or her turn to answer a question and how long he or she has to answer it. More than two politicians can take part in the debate by splitting the time to answer questions equally among them. The more politicians that are allowed to debate, the less time each politician is allowed to speak, assuming the debate has a time limit. The politicians should address their responses to the person asking the questions. If they address their responses to each other, they are not really operating in TDMA mode, but are using time division duplexing (TDD) to talk to each other. If some kind of multiple access technique were not used in a political debate, the debate would likely digress into a babble of nonsense that no one would understand as each politician tried to talk over the other one. Of course, there is a school of thought that believes any time a politician speaks it is a babble of nonsense.

Figure 2.1 shows how TDMA is used in Digital PCS to increase the capacity of the cellular system over that achievable by AMPS. In Digital PCS, TDMA is used to allow from one to six users to share a single

Advantages of Digital PCS   21

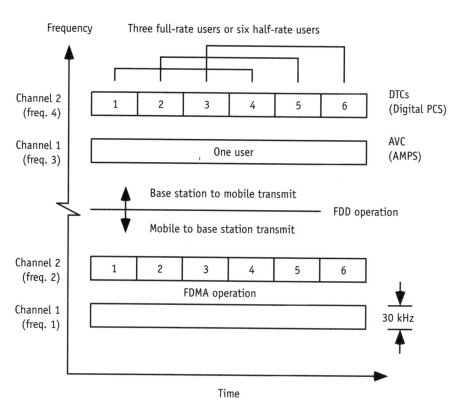

Figure 2.1 Digital PCS capacity increase with TDMA.

30-kHz RF channel. This is accomplished by splitting the channel into six equal intervals of time, or time slots. In the typical mode of operation for Digital PCS, called full rate, each user is allowed access to two of the six time slots, equally spaced in time. This means that one 30-kHz RF channel can serve three simultaneous full-rate users, increasing the instantaneous capacity by a factor of three over AMPS. Digital PCS actually allows for each of the six time slots to be assigned to different users in what is called half-rate mode. In half-rate mode, the capacity of the air interface is increased by a factor of six over AMPS. However, no half-rate vocoder has been developed for Digital PCS. If a cell was assigned only one RF channel, only three full-rate users at a time could be supported over the air interface. FDMA is used in Digital PCS to support more users by assigning multiple RF channels for use in the same cell, just like in

AMPS. A mobile user sends and receives on two different frequencies, which is called frequency division duplexing (FDD). So, in reality, Digital PCS is a cellular radiotelephone system that uses a combination of TDMA and FDMA as the multiple access techniques and FDD as the duplexing technique.

The capacity increase of Digital PCS over AMPS due to the use of TDMA can be more than a factor of three for two reasons. First, the control channel in a Digital PCS system (a DCCH) only occupies 1/3 of an RF channel (two time slots out of six), while the control channel in an AMPS system (an ACC) occupies a full RF channel. For a pure Digital PCS system—that is, one that does not contain ACCs for backward compatibility to AMPS mobiles—the instantaneous capacity increase over AMPS is given by

$$\frac{3N_C - 1}{N_C - 1} \qquad (2.1)$$

where N_C is the number of RF channels per cell or sector, and must be greater than 1 for an AMPS system. For a large number of RF channels, the instantaneous capacity increase is approximately 3. For a small number of RF channels, however, the instantaneous capacity increase is much greater. This is significant for picocells, in which very few RF channels may be present. It is worth noting that one control channel per sector is assumed in (2.1).

Trunking efficiency is the second and biggest reason why TDMA allows Digital PCS to provide more than a factor of three increase in capacity over AMPS. The scope of this text does not include a thorough treatment of trunking theory, so this discussion will be limited to the basic concept. The term *instantaneous capacity* has been used in the previous discussion to indicate the number of simultaneous users that can be accommodated. Trunking is used in cellular radio telephone systems to accommodate more users than the instantaneous capacity allows. With trunking, fewer resources are available than there are potential users, so there is some probability, called the probability of blocking, that a user will be unable to access the system at a given time. Cellular radio telephone systems are typically engineered to provide a low probability of blocking during the busy hour. The probability of blocking during the busy hour is called the grade of service, and a 2% grade of service is typically the target for system designers.

There are different models that are used to calculate the traffic-carrying capacity of a system designed for a particular grade of service. Most of these models calculate the traffic intensity in erlangs for a given number of channels and a particular grade of service. An estimate of the average erlangs per users is multiplied by the number of erlangs available in a sector to determine the user capacity of the sector. The most popular model for estimating cellular system capacity is the Erlang B model, which assumes that all blocked calls are cleared and not queued. Table 2.1 reproduces part of the Erlang B model for a 2% grade of service.

Using (2.1), if there are 15 RF channels available in a sector of a cell, then for that sector there are 44 traffic-carrying channels in a Digital PCS deployment and 14 traffic-carrying channels in an AMPS deployment. From Table 2.1, this equates to 34.7 erlangs for a Digital PCS system, and 8.2 erlangs for an AMPS system, for a 2% grade of service and an Erlang B traffic model. For this example, the Digital PCS system offers a capacity increase of $34.7/8.2 = 4.23$ times that of AMPS.

Table 2.1
Erlang B Traffic Model for 2% Grade of Service (N_T = Number of Traffic Channels, A = Traffic Intensity in Erlangs)

N_T	A	N_T	A	N_T	A	N_T	A	N_T	A
1	0.0204	21	14.0	41	31.9	61	50.6	81	69.6
2	0.223	22	14.9	42	32.8	62	51.5	82	70.6
3	0.602	23	15.8	43	33.8	63	52.5	83	71.6
4	1.09	24	16.6	44	34.7	64	53.4	84	72.5
5	1.66	25	17.5	45	35.6	65	54.4	85	73.5
6	2.28	26	18.4	46	36.5	66	55.3	86	74.5
7	2.94	27	19.3	47	37.5	67	56.3	87	75.4
8	3.63	28	20.2	48	38.4	68	57.2	88	76.4
9	4.34	29	21.0	49	39.3	69	58.2	89	77.3
10	5.08	30	21.9	50	40.3	70	59.1	90	78.3
11	5.84	31	22.8	51	41.2	71	60.1	91	79.3
12	6.61	32	23.7	52	42.1	72	61.0	92	80.2

Table 2.1 (continued)

N_T	A	N_T	A	N_T	A	N_T	A	N_T	A
13	7.40	33	24.6	53	43.1	73	62.0	93	81.2
14	8.20	34	25.5	54	44.0	74	62.9	94	82.2
15	9.01	35	26.4	55	44.9	75	63.9	95	83.1
16	9.83	36	27.3	56	45.9	76	64.9	96	84.1
17	10.7	37	28.3	57	46.8	77	65.8	97	85.1
18	11.5	38	29.2	58	47.8	78	66.8	98	86.0
19	12.3	39	30.1	59	48.7	79	67.7	99	87.0
20	13.2	40	31.0	60	49.6	80	68.7	100	88.0

2.1.2 Hierarchical cell structures

HCS is a method of increasing the capacity of a cellular system by forming layers of service [1]. The example of the political debate described earlier in this chapter can be expanded to describe how HCS works. Suppose the political debate were between presidential and vice presidential candidates. Having both types of candidates in the same debate would be difficult for two reasons. First, there would probably not be enough time for everyone to speak due to the sheer number of people. Second, the presidential candidates would tend to dominate the conversation, and the vice presidential candidates might not get heard. One way to solve this problem would be for the vice presidential candidates to debate on the ground floor while the presidential candidates debate on the second floor. There would be two debates going on simultaneously, and all the candidates would have enough time to speak. This sounds like a good building to stay away from, though.

HCS is implemented in Digital PCS through the use of tools that direct a mobile to the most appropriate cell site from which to obtain service, be it a macrocell, a microcell, or a picocell. These tools are not available in AMPS. This layered HCS architecture is illustrated in Figure 2.2. One might think of the macrocells as the second floor where the presidential candidates debate, and the microcells as the first floor where the vice presidential candidates debate. A macrocell is typically implemented as an umbrella cell over multiple microcells. Similarly, a microcell can be implemented as an umbrella cell to one or more

Advantages of Digital PCS 25

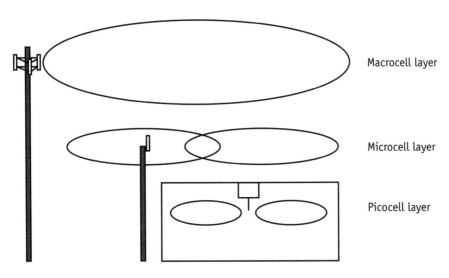

Figure 2.2 Example of HCS in a Digital PCS system.

picocells. Picocells are typically implemented for use inside buildings. Mobile users in fast-moving vehicles are typically connected to macrocells. Slow-moving mobile traffic and outdoor pedestrian users typically are connected to microcells, while pedestrians and stationary users inside buildings are connected to picocells. The tools used to direct a mobile to the most appropriate cell site include cell type identifiers, signal strength biases, time delay requirements, and mobile capabilities. Chapter 9 describes in detail how these tools are used to implement HCS and direct mobile traffic as desired by the service provider.

It may not be obvious how HCS is used in a Digital PCS cellular system to increase capacity. The capacity increase is enabled by distributing traffic among cells and by the tighter frequency reuse that is available in each lower layer of cells in a given geographic area. The tighter frequency reuse is made possible by the higher path loss experienced in the microcell and picocell environments than in the macrocellular environment. The higher path loss is the result of additional clutter—for example, buildings, terrain, and foliage—experienced at lower base station antenna heights. An example will help to show the advantages of HCS for capacity increase. Only macrocells and microcells are considered in the example. Suppose 30 MHz of spectrum is allocated for use in a cellular system. For both AMPS and Digital PCS systems, half of this spectrum would be used in the forward direction and half in the reverse direction, because both systems use FDD. There is therefore 15 MHz of spectrum that can be split into smaller RF channels. Both AMPS and Digital PCS use 30-kHz channels, so 500 RF channels are available for both systems.

The capacity of a Digital PCS system and an AMPS system using the available 15 MHz of spectrum can be compared first by assuming that all cells are macrocells. A typical frequency reuse pattern for Digital PCS and AMPS macrocells is 7/21. That is, frequencies are reused every seventh cell, and each cell is divided into three sectors. This frequency reuse pattern results in $7 \times 3 = 21$ channel sets and has been demonstrated to meet the C/I requirements for good-quality voice service under most radio propagation conditions. With 21 channel sets and 500 RF channels, each sector of a macrocell can be allocated 23 RF channels. There are therefore 68 traffic channels per sector for a Digital PCS deployment, and 22 traffic channels per sector for an AMPS deployment after control channels are removed. From Table 2.1, this corresponds to 57.2 and 14.9 erlangs per sector for a Digital PCS deployment and an AMPS deployment, respectively. For this example, the capacity increase of Digital PCS over AMPS for an all-macrocell deployment in 15 MHz of spectrum with a 7/21 frequency reuse pattern is a factor of $57.2/14.9 = 3.84$.

With HCS, the capacity increases even more dramatically. Assume enough channels are allocated for macrocell use in the Digital PCS deployment to match the capacity that would have been provided by an AMPS macrocell deployment, and the rest of the channels are used for a microcell underlay. For the macrocellular deployment, 8 RF channels per sector are required (corresponding to 23 traffic channels and 1 control channel), or 168 channels total with a 7/21 frequency reuse pattern. There are still $500 - 168 = 332$ RF channels left for the microcell underlay. Now assume the microcells are omnidirectional and can be implemented with a frequency reuse pattern of 12 due to the higher propagation loss encountered in the microcell environment. There are therefore $332/12 = 27$ RF channels available for use in each microcell, which corresponds to 80 traffic channels and 1 control channel per microcell. If one microcell covers an area equivalent to one sector of a macrocell, then within this area there is a total of 23 traffic channels

(15.8 erlangs) for the macrocell and 80 traffic channels (68.7 erlangs) for the microcell, or 84.5 erlangs total. For this example, the capacity increase of Digital PCS with HCS over AMPS is a factor of 84.5/14.9 = 5.67. Figure 2.3 summarizes the capacity increase with HCS.

2.1.3 Adaptive channel allocation

ACA is a method of increasing the capacity of a cellular system while maintaining the engineered grade of service by automatically and dynamically changing the frequency reuse pattern to optimally match the propagation environment and traffic distribution [2]. With ACA, the

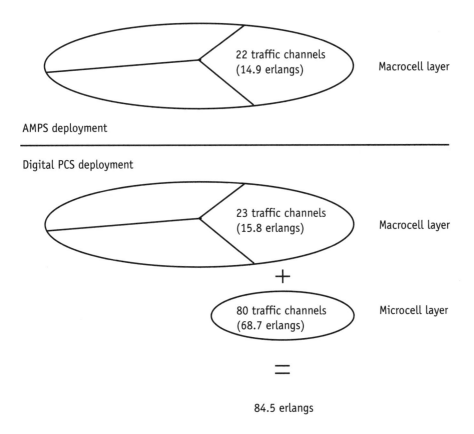

Figure 2.3 Capacity increase with HCS.

number of RF channels available in any given sector may increase or decrease over time, depending upon the need in that sector and other sectors. This is important because not all sectors must handle the same amount of traffic. For example, a sector of a cell that serves the intersection of two busy highways likely carries more traffic than another sector of the same cell that covers a residential area. ACA might be thought of as the ultimate RF engineer, with accurate knowledge of the C/I environment and traffic patterns in every sector at every point in time.

Digital PCS enables ACA through the mobile assisted channel allocation (MACA) procedure. With MACA, the network requests mobiles operating on a DCCH to measure and report the signal strength on up to 15 channels in addition to the current DCCH. MACA is described in greater detail in Chapter 8. The MACA report provides the network with downlink signal strength measurements on channels that may be in use in other sectors. The network may combine these downlink signal strength measurements with uplink measurements that are made at the base station to form a composite view of the availability of channels for use in a sector. If the signal strength of a channel measured on both the uplink and downlink within a sector is below a threshold, the network may identify this channel as available for use within that sector. With 136+, the mobiles have the additional ability to estimate and report C/I on channels, which provides even more information for the ACA process.

ACA can be combined with HCS to optimize the RF channel assignment between layers of cells. ACA can even make it possible to reuse RF channels between the layers of cells. In the example of HCS described previously, the spectrum was segmented into channels used for the macrocell layer and channels used for microcells. If 100% frequency sharing were realized with ACA, all 500 channels would be available for assignment in both cell layers. Per macrocell sector, there would be 68 traffic channels (57.2 erlangs), as calculated previously. Per microcell, there would be 122 traffic channels (109.4 erlangs, not shown in Table 2.1) with a frequency reuse pattern of 12. For this example, the capacity increase of Digital PCS over AMPS for an HCS deployment with ACA is a factor of $(57.2 + 109.4)/14.9 = 11.18$. Figure 2.4 summarizes the capacity increase with ACA.

Advantages of Digital PCS 29

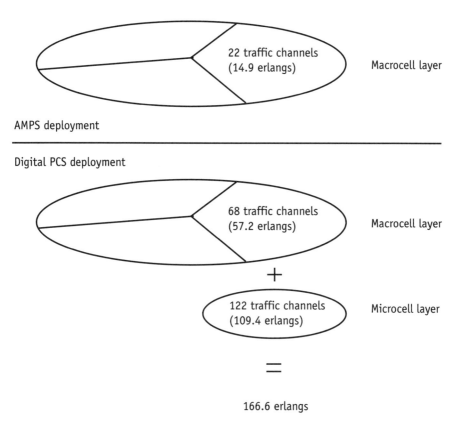

Figure 2.4 Capacity increase with ACA.

2.2 Feature flexibility

2.2.1 Layered protocol

The layered protocol of the Digital PCS standard makes it possible for new features to be added without impacting lower layers or causing incompatibility with mobiles or networks that do not support the new features [3]. New features can be implemented more rapidly if the impact of the new feature on the current implementation is minimized. For example, if a new feature requires changes to both the software and hardware of a mobile, it will undoubtedly take longer to develop and test than a change that impacts only the mobile software. Even if a new feature only impacts the mobile software, it can still lengthen the development and testing cycles if the changes touch many different parts of the code, or if the feature requires new interaction between the mobile and the network. The importance of adequate testing of new features should not be underestimated. Regression testing must be conducted to ensure that any changes to add a new feature have not "broken" some other feature or functionality, and interoperability testing must be conducted to ensure that both the network and the mobile have implemented a new feature in the same way. Testing is explored in more detail in Chapter 18.

The Digital PCS standard separates the air interface into a three-layered protocol. Layer 1 is the physical layer and defines the RF channel bandwidth, power output and modulation characteristics, time slot and frame structure, forward error correction, and interleaving/deinterleaving. Chapter 4 of this text is devoted to a description of the Digital PCS physical layer. Layer 2 is the data link layer and defines the addressing, frame delimiting, rules for segmenting and reassembling layer 3 messages requiring multiple time slots for transmission, generation of cyclic redundancy check (CRC), rules for mobile access to channels, and retransmission control. Chapter 5 of this text is devoted to a description of layer 2. Layer 3 is the network layer and defines the mobile states and processes and the message set used between the mobile and the base station. Chapters 6 and 8 of this text describe layer 3 of the Digital PCS standard.

Most new features are added to Digital PCS through additions to layer 3 only. This means that layer 1, which is mostly implemented in hardware, and layer 2, which is mostly implemented in digital signal processor (DSP) code, can remain unchanged when a new feature is added. For the case of a mobile, many new features are added via new software in the microcontroller to control the signaling with the base station and the interface to the mobile user. The structure of layer 3 messages simplifies the addition of new features. Every layer 3 message is composed of separate information elements. The first information element is always the Protocol Discriminator, which identifies the standard that defines the message. The second information element is the Message Type, which defines how to interpret the remaining information elements. The Protocol Discriminator and Message Type are mandatory information elements that must be present in every layer 3 message. Any other mandatory information elements are included immediately after these two. The length of all mandatory information elements is defined by the Digital PCS standard, so every mobile or base station knows how many bits to associate with every received mandatory information element in a received layer 3 message. Optional information elements may be included in a layer 3 message after the mandatory information elements. Every optional information element contains a remaining length field as the first set of bits. This tells the mobile or base station how many bits are associated with the optional information element.

The structured layer 3 protocol just described makes it easy to implement new features that require signaling between the mobile and base station. If a new feature requires information to be exchanged between the mobile and base station, new information elements can be added to existing layer 3 messages, or new layer 3 messages can be defined. Mobiles or base stations that have not implemented the new feature simply ignore the new information elements or new messages. An example will help to illustrate this point. Suppose a service provider desires to implement a new feature to display the current temperature on users' mobiles. The temperature measurement is sent to the MSC and ultimately to the base stations connected to the MSC. The temperature measurement is included in a new optional information element that is added to an existing message on the broadcast control channel of the DCCH. Mobiles that were placed in the field prior to the implementation of the new feature simply ignore the new information element and continue to read messages they understand and operate as before. However, new mobiles that have implemented software to support this new feature can read the broadcast information and obtain the current temperature from the new optional information element. The mobile can then either store the information for display upon user request or continuously display the temperature, depending upon the user interface capabilities of the mobile and the requirements of the service provider.

2.2.2 Teleservices

A teleservice was defined in Chapter 1 as an application that uses the air interface and network interface as the bearers for transport between a teleservice server and a mobile. Essentially, a teleservice uses the cellular network as a pipe through which a feature is provided to a user. The new feature is developed on the teleservice server (perhaps by the service provider using a service creation environment) and the mobile, but little or no development is needed elsewhere in the cellular network to support the feature. This can shorten the time from the definition of a feature to commercial launch because new network functionality often requires the most development and testing time.

A teleservice resides above layer 3 of the Digital PCS standard. That is, a teleservice is an application that is encapsulated in a layer 3 message, the R-DATA message. TIA/EIA-41 defines a similar layer 3 message, called Short Message Delivery–Point-to-Point (SMDPP), within which a teleservice resides. A higher layer protocol identifier (HLPI) identifies the teleservice to the mobile, and a teleservice ID identifies the teleservice to the teleservice server. There is a one-to-one mapping of teleservice IDs to HLPIs. For most teleservices, the cellular network simply takes SMDPP messages it receives from a teleservice server and reformats the message as an R-DATA message for transmission over the air interface, and vice versa. The cellular network is typically not required to read or understand the information contained in the SMDPP or R-DATA messages.

Providing the current temperature to a mobile user is an example of a feature that could be implemented as a teleservice instead of a new message or information element on the DCCH. Instead of sending the temperature measurement to the MSC, the measurement is sent to the teleservice server. The teleservice server formats the temperature measurement into a teleservice specified by the teleservice ID, places the teleservice within an SMDPP message, and sends the SMDPP message to the MSC. The MSC and base station reformat the SMDPP message into an R-DATA message with the appropriate HLPI and deliver it to the mobile. The mobile knows how to interpret the R-DATA message because it recognizes the HLPI for the temperature measurement. This is an example of a teleservice that could be provided as a broadcast teleservice instead of a point-to-point teleservice, so that many mobiles may receive the message at the same time. More signaling is actually required to deliver a teleservice from a teleservice server to a mobile, which will be described in a subsequent chapter.

2.3 AMPS compatibility

Two important aspects of Digital PCS for service providers that have installed AMPS networks is the ability for Digital PCS to coexist with AMPS, and to provide a migration path to digital service [4]. Both AMPS and Digital PCS use 30-kHz channels and can be deployed with the same frequency reuse plan, as discussed earlier in this chapter. Furthermore, Digital PCS defines dual-mode operation, which includes the entire AMPS functionality as a subset of the standard. Finally, the DCCH used for signaling and control in Digital PCS systems can be located anywhere in the allocated spectrum.

AMPS mobiles can be accommodated in a Digital PCS network operating in the 800-MHz cellular band by broadcasting ACCs and assigning AVCs, in addition to broadcasting DCCHs and assigning DTCs. An AMPS mobile operating in a Digital PCS network will scan for ACCs, be assigned to AVCs, and perform handoffs exactly as described in Chapter 1 for an AMPS network. However, a Digital PCS network not requiring support for AMPS mobiles or a network operating in the 1,900-MHz PCS band does not need to broadcast ACCs or assign AVCs.

A Digital PCS mobile operating in a Digital PCS network that supports AMPS may scan for ACCs or DCCHs. The preferred mode of operation is for the Digital PCS mobile to camp on a DCCH for signaling and control, instead of an ACC. This allows the mobile to take advantage of the features available on the DCCH, including sleep mode, HCS, and teleservices. The Digital PCS mobile may be assigned to an AVC or a DTC, and may be handed off from one type of channel to another. The preferred mode of operation is for the mobile to operate on the DTC, to take advantage of the features available on the DTC such as voice privacy and increased talk time. It is worth noting that since all Digital PCS mobiles are dual-mode, they can also operate on AMPS networks just like AMPS mobiles.

An AMPS network can be migrated to a Digital PCS network over time. Initially, both AMPS and Digital PCS service can be supported. The DCCHs can be placed anywhere in the allocated spectrum and a pointer to a DCCH can be placed on every ACC to assist Digital PCS mobiles in finding DCCHs. Upon call release from an AVC, DCCH location information can be provided to a Digital PCS mobile to aid the mobile in finding the most appropriate DCCH. DTCs can be located on any 30-kHz RF channel in the allocated spectrum. As users are migrated to Digital PCS, more 30-kHz channels can be converted from AVCs to DTCs. The conversion from AVCs to DTCs should be closely tied to the number of AMPS mobiles homed to the network and the number of roaming AMPS mobiles expected in the network. Ultimately, a minimal number of AVCs and the entire set of 21 ACCs should be retained in the network to support roaming mobiles, even when the entire base of home mobiles are converted from AMPS to Digital PCS.

References

[1] Post, A. J., and R. Djurkovic, "Small Cells, Big Returns," *Wireless Review*, Vol. 15, No. 12, June 15, 1998, pp. 30–39.

[2] Katzela, I., and M. Naghshineh, "Channel Assignment Schemes for Cellular Mobile Telecommunication Systems: A Comprehensive Survey," *IEEE Personal Communications*, Vol. 3, No. 3, June 1996, pp. 10–31.

[3] Jain, B. N., and A. K. Agrawala, *Open Systems Interconnection: Its Architecture and Protocols*, 3d ed., New York: McGraw-Hill, 1993, pp. 15–21.

[4] Smith, C., *Practical Cellular and PCS Design*, New York: McGraw-Hill, 1997, pp. 2.25–2.28.

3

The Digital PCS Family of Standards

DIGITAL PCS IS defined by a family of voluntary standards that establish the compatibility and performance requirements for the air interface technology. Standards assist manufacturers, service providers, and end users to build, buy, and use products that provide mutually agreed upon features, functionality, and performance. Standards also facilitate the interoperability of products built by different manufacturers. With standards, the service provider and end user have more choices in products than would otherwise be the case, and manufacturers have access to more markets than otherwise. The Digital PCS family of standards includes IS-136 (now TIA/EIA-136), IS-137, IS-138, IS-641, IS-130, and IS-135, which this chapter summarizes. Eventually, all these standards will become parts of the TIA/EIA-136 standard.

3.1 Telecommunications Industry Association

The Telecommunications Industry Association (TIA) developed the Digital PCS family of standards. The TIA is a trade organization that operates in association with the Electronics Industry Association (EIA) to develop telecommunications standards for North America. Engineering committees of the TIA formulate standards through the efforts of both TIA member companies and nonmember participating companies. The TIA designates its engineering committees as TR committees. This is a holdover from the first TIA engineering committee, TR-8, originally known as the Transmitter Division (hence the letters *TR* from *transmitter*). TR-45, the Mobile and Personal Communications Standards engineering committee, develops the majority of cellular standards for North America. These include air interface standards such as EIA/TIA-553, IS-54, and TIA/EIA-136, and intersystem standards such as TIA/EIA-41. Different engineering subcommittees formulate standards for differing air interface technologies and the intersystem interface. Within TR-45, engineering subcommittee TR-45.3, Time Division Digital Cellular Technology, maintains the Digital PCS air interface standards. Subcommittee TR-45.2, Cellular Intersystem Operation, maintains the TIA/EIA-41 standard for intersystem signaling used with Digital PCS.

TR-45.3 is organized into working groups focused on different aspects of the Digital PCS standards, as shown in Table 3.1. Working Group 2 is responsible for data services, including IS-130 and IS-135 circuit-switched data and 136+ packet data. Speech services are under the purview of Working Group 5. Working Group 5 standardizes vocoders such as IS-641. Working Group 6 is responsible for all aspects of the DCCH, as well as the signaling on the TIA/EIA-136 control and traffic channels. Each working group works and meets independently of each other and coordinates their activities at the engineering subcommittee level.

Working groups are contribution-driven. A contribution is a written recommendation from a working group participant, with background material to justify the recommendation. The working group must reach consensus on the recommendation for it to be adopted. A new standard or a revision to a standard normally begins with a contribution from one

Table 3.1
TR-45.3 Working Groups

TIA TR-45.3 Working Groups	Responsibility	Example of Standard Developed
TR-45.3.2	Data services	IS-130/135
TR-45.3.5	Speech services	IS-641
TR-45.3.6	Digital control channel	TIA/EIA-136

or more participants in TIA. The development of a new feature or service within a standard also begins with a contribution.

There are three stages to the development of standards text for new features or services, called by the somewhat uninspired names of stage 1, stage 2, and stage 3. Stage 1 is a description of a feature or service from the user's point of view. It describes what the proposed feature or service will do for the user. Stage 2 is a description of a feature or service as implemented in the network. It describes the interaction of different elements in the cellular network to provide the feature or service to the end user. Stage 3 is a detailed description of the procedures, protocols, physical characteristics, and performance required to provide the feature or service to the end user. It is the stage 3 description that becomes standards text. Stage 1 should be completed before stage 2, and stage 2 before stage 3. This guarantees thorough definition of a new feature or service and increases the likelihood of its implementation.

Approved stage 3 text is added to the adopted baseline text for the standard. The baseline text is frozen when the working group determines by consensus that all required new material has been approved. Verification and validation (V&V) comes next, in which the proposed standards text is reviewed for technical and editorial accuracy. Discrepancies in the proposed standard are identified and corrected. The V&V process is an internal review conducted before submitting the proposed standard for approval. The proposed standard is submitted for ballot once the V&V process is complete.

3.2 Interim Standards and American National Standards

A TIA standard may be published as an Interim Standard when time is of the essence due to an urgent industry need for a technology. The TIA releases an Interim Standard for industry use for a limited period of time. It may eventually become an American National Standard if balloted and approved by the American National Standards Institute (ANSI). A TIA Interim Standard is identified by the designator TIA/EIA/IS, followed by a hyphen and a numeral. When an Interim Standard becomes an American National Standard, the designator is changed to ANSI/TIA/EIA and the IS designator is dropped. Often, the TIA/EIA designator is left off when discussing an Interim Standard—for example, IS-136. Similarly, the ANSI designator is often left off when discussing a national standard. For example, the IS-54 standard has become an American National Standard with the designation TIA/EIA-627. Most people still know this standard as IS-54, however. The AMPS standard EIA/TIA-553 is an American National Standard that was formulated before the TIA was spun-off from the EIA, hence the acronyms are reversed.

After a proposed Interim Standard is formulated, a letter ballot is distributed to TIA member organizations. Organizations may vote yes, yes with comments, or no with comments. The engineering subcommittee attempts to resolve any negative ballot comments received, and reballots the proposed standard if technically substantive change result from comment resolution. A default ballot may be issued if limited technical changes have been made to the proposed standard to resolve a negative ballot comment. The default ballot results must show a consensus in favor of adoption of the proposed standard. If a ballot results in adoption of the proposed standard, the TIA confirms that it has been prepared according to guidelines and publishes it as an Interim Standard.

The value of an Interim Standard is the rapid approval process, primarily due to the limited requirements for ballot review. However, an Interim Standard only has a lifetime of three years. It behooves the TIA formulating body to eventually convert the Interim Standard to an American National Standard by presenting it for approval by ANSI. ANSI requires a broader group of ballot reviewers, a longer ballot cycle, and a more formal ballot resolution process.

3.3 The Digital PCS standards

The Digital PCS family of standards, summarized in Table 3.2, originated as Interim Standards in TIA Engineering subcommittee TR45.3. IS-136 grew out of the IS-54 standard and defines the control, traffic, and voice channels used in Digital PCS. IS-137 and IS-138 specify the minimum performance requirements for Digital PCS mobiles and base stations, respectively. IS-641 defines a high-quality vocoder used in Digital PCS. The circuit-switched data standards for Digital PCS are defined by IS-130 and IS-135.

3.3.1 IS-136

IS-136 (now TIA/EIA-136) describes the attributes of the channels used in Digital PCS, from the physical layer through the network layer. By far, the largest section of the standard is the part describing the DCCH, undoubtedly because the DCCH is the most complex of the Digital PCS channels. The definition of the DCCH is also the primary difference between IS-136 and IS-54, apart from the new vocoder defined by IS-641. Because IS-136 builds on IS-54, most of the standards text describing the ACC, AVC, and DTC remains in the IS-54 format. The format of the standards text describing the DCCH is much different, however. Three major sections of IS-136 describe the DCCH, with each section devoted to one of the layers of the IS-136 protocol. This makes it easier for implementers of the standard to focus on the parts they need to understand. For example, an RF designer needs to focus on the physical layer of IS-136. The link layer of IS-136 is the major focus of a DSP designer. Call processing and user interface software designers concentrate on the network layer.

The first version of IS-136, called Revision 0, simply adds the DCCH to the IS-54 control, voice, and traffic channels. The new features in Revision 0 of IS-136 include sleep mode, short message service, and nonpublic service. Revision A to IS-136 adds dual-band 800-MHz and 1,900-MHz support, call setup and handoff for the IS-641 vocoder, over-the-air activation, calling name presentation, and support for full-rate asynchronous data and fax. 136+ is the next revision to IS-136 and adds additional teleservice capability, intelligent roaming, a more robust and higher fidelity voice service, and packet data.

Table 3.2
Summary of the Digital PCS Family of Standards

Designation	Title	Content
IS-136	TDMA Cellular/PCS—Radio Interface—Mobile Station-Base Station Compatibility Standard	Air interface requirements for the digital control channel, analog control channel, digital traffic channel, and analog voice channel used in Digital PCS
IS-137	TDMA Cellular/PCS—Radio Interface—Minimum Performance Standards for Mobile Stations	Minimum performance requirements for 800-MHz and 1,900-MHz Digital PCS mobile stations
IS-138	TDMA Cellular/PCS—Radio Interface—Minimum Performance Standards for Base Stations	Minimum performance requirements for 800-MHz and 1,900-MHz Digital PCS base stations
IS-641	TDMA Cellular PCS—Radio Interface—Enhanced Full-Rate Speech Codec	Description of a 7.4-Kbps ACELP vocoder and associated forward error correction for use with Digital PCS
IS-130	TDMA Cellular/PCS—Radio Interface—Radio Link Protocol 1	Definition of a radio link protocol used to asynchronously transport data across an IS-136 radio interface
IS-135	TDMA Cellular/PCS—TDMA Services—Async Data and Fax	Definition of the network protocol used to asynchronously transport data across and IS-136 radio interface

IS-136 Revision A is the same as ANSI/TIA/EIA-136, and 136+ is the same as ANSI/TIA/EIA-136-A. This might appear confusing, but only because of the timing associated with converting IS-136 to an American National Standard. IS-136 Revision A was the most recently adopted version of IS-136 when the three-year time limit on an Interim Standard expired. Because 136+ was not ready for ballot, TR-45.3 chose to submit IS-136 Revision A for ANSI ballot. All revisions to the standard after the adoption of ANSI/TIA/EIA-136, including 136+, are subject to the full ANSI ballot process and become revisions to the ANSI standard.

3.3.2 IS-137 and IS-138

IS-137 defines the minimum performance requirements for Digital PCS mobiles operating at 800 MHz and at 1,900 MHz. These minimum performance requirements fall into three categories: receiver, transmitter, and environmental. For each minimum requirement, the requirement is defined, the method of measurement is presented, and the minimum standard is identified. IS-137 includes receiver minimum standards for frequency coverage and acquisition time, demodulation, receive audio frequency response, receiver sensitivity, adjacent and alternate channel desensitization, protection against spurious response, bit error rate, protection against multipath, and signal strength measurement accuracy. The standard includes transmitter minimum requirements for frequency stability, carrier switching time, power output, power transition time, modulation type and stability, transmit audio frequency response, limitation of emissions, and time alignment. The environmental requirements for the mobile in terms of temperature, power supply voltage, humidity, vibration, and shock are also defined in IS-137.

The companion standard to IS-137 for Digital PCS base stations is IS-138. IS-138 contains base station receiver, transmitter, and environmental requirements, similar to IS-137. Environmental requirements for the base station are only included for temperature, power supply voltage, and humidity. Unlike mobiles, vibration and shock requirements are not defined for base stations because they are typically shielded from these conditions.

IS-137 and IS-138 are supporting standards to IS-136. They provide mobile and base station manufacturers with minimum performance requirements for the implementation of IS-136. A physical or link layer change to IS-136 often necessitates a change to IS-137 and IS-138. IS-137 Revision 0 and IS-138 Revision 0 only contain requirements for 800-MHz operation. Revision A to these standards adds requirements for 1,900-MHz mobile and base station operation. Minor changes are also made in Revision A of IS-137 and IS-138 to accommodate the use of the IS-641 vocoder.

3.3.3 IS-641

IS-641 describes the ACELP vocoder that replaces the VSELP vocoder used in IS-54. A detailed description of the IS-641 voice encoder and decoder is provided in the standard, along with a description of the channel encoder and decoder used to provide error protection to the IS-641 coded voice. An example bad frame masking algorithm is also described in IS-641. This algorithm describes an example of the processing that can be used on decoded speech to make it more intelligible when speech frames have been corrupted by RF channel impairments. IS-641 also describes a bit exact software representation of the voice encoder and decoder, which must be used in Digital PCS mobiles. This guarantees that all Digital PCS mobile stations operate on user speech the same way. Base stations are not required to use a bit exact form of the IS-641 vocoder, however. Test vectors are included for the bit exact representation of the voice encoder and decoder to ensure that the standard is implemented as specified.

IS-641 is another supporting standard to IS-136. IS-136 defines the physical layer that the IS-641 vocoder uses. IS-136 also defines the protocol used between the mobile and base station to set up calls and perform handoffs with the IS-641 vocoder. Revision A of IS-641 adds *discontinuous transmission* capability to the vocoder. With discontinuous transmission, voice activity detection is used to identify pauses in user speech, and only actual speech samples are coded and transmitted. This differs from typical vocoder implementations in which pauses in user speech are processed and transmitted as if they were actual speech segments. Discontinuous transmission allows the mobile to turn off its transmitter for a longer period of time during a conversation, thereby decreasing current drain and increasing talk time. Also added in Revision A to IS-641 is *comfort noise*. Comfort noise parameters can be transmitted from the voice encoder to fill in the silent gaps caused by discontinuous transmission. This can increase the perceived quality of the conversation.

3.3.4 IS-130 and IS-135

IS-130 and IS-135 are companion standards that require portions of IS-136 to be implemented for them to be used. IS-130 is the link layer that is used with the IS-135 network layer to provide asynchronous data and fax service on an IS-136 digital traffic channel. IS-130 defines a radio link protocol, RLP1, which includes link establishment, link supervision, acknowledged data transport, unacknowledged data transport, data qualification, data compression, encryption, and flow control. IS-135 defines a network protocol for asynchronous data and fax transport that includes call setup, supervision, clearing, AT command handling, user data transport, online command signaling, break signaling, and signaling leads for the service. IS-135 requires a link layer protocol such as IS-130 and uses the IS-136 call control for call setup, supervision, and clearing. With IS-130 and IS-135, asynchronous data and fax service can be provided over a half-, full-, double-, or triple-rate digital traffic channel. This allows for a range of data rates based on user need and mobile and system capabilities. IS-130 and IS-135 are the subjects of Chapter 13.

3.3.5 TIA/EIA-136

The American National Standard TIA/EIA-136 will merge the Digital PCS family of standards into a single standard with multiple parts. Working Group 6 of TR-45.3 restructured IS-136 and incorporated the remaining family of Digital PCS standards into one to make the Digital PCS standard easier to use, track, and revise. Table 3.3 summarizes the parts of TIA/EIA-136. The benefits of the multipart standard can be observed by considering the TIA/EIA-136-7XX parts, which define the Digital PCS teleservices. Under TIA/EIA-136, a new teleservice can be developed and the standard for that teleservice can be balloted without requiring the balloting of any other parts of the Digital PCS standard. Mobile and teleservice server software developers can pull out this small part of the standard and develop the software to support the teleservice without having to wade through information not of use to them for the task at hand.

Table 3.3
TIA/EIA-136 Parts

Part Number	Title
TIA/EIA-136-000	List of Parts
TIA/EIA-136-0XX	Miscellaneous Information
TIA/EIA-136-009	Introduction
TIA/EIA-136-010	Optional Mobile Station Facilities
TIA/EIA-136-020	SOC, BSMC, and Other Codes
TIA/EIA-136-1XX	Channels
TIA/EIA-136-100	Introduction to Channels
TIA/EIA-136-110	RF Channel Assignments
TIA/EIA-136-121	Digital Control Channel Layer 1
TIA/EIA-136-122	Digital Control Channel Layer 2
TIA/EIA-136-123	Digital Control Channel Layer 3
TIA/EIA-136-131	Digital Traffic Channel Layer 1
TIA/EIA-136-132	Digital Traffic Channel Layer 2
TIA/EIA-136-133	Digital Traffic Channel Layer 3
TIA/EIA-136-140	Analog Control Channel
TIA/EIA-136-150	Analog Voice Channel
TIA/EIA-136-2XX	Minimum Performance
TIA/EIA-136-200	Introduction
TIA/EIA-136-210	ACELP Minimum Performance
TIA/EIA-136-220	VSELP Minimum Performance
TIA/EIA-136-270	Mobile Stations Minimum Performance
TIA/EIA-136-280	Base Stations Minimum Performance
TIA/EIA-136-3XX	Data Services
TIA/EIA-136-300	Introduction
TIA/EIA-136-310	Radio Link Protocol-1
TIA/EIA-136-320	Radio Link Protocol-2
TIA/EIA-136-330	Packet Data
TIA/EIA-136-350	Async Data/Fax
TIA/EIA-136-4XX	Vocoders
TIA/EIA-136-400	Introduction
TIA/EIA-136-410	ACELP
TIA/EIA-136-420	VSELP
TIA/EIA-136-430	US1
TIA/EIA-136-5XX	Security
TIA/EIA-136-500	Introduction

The Digital PCS Family of Standards 45

Part Number	Title
TIA/EIA-136-510	Authentication, Encryption of Signaling Information/User Data, and Privacy
TIA/EIA-136-511	Messages Subject to Encryption
TIA/EIA-136-6XX	Teleservice Transport
TIA/EIA-136-600	Introduction
TIA/EIA-136-610	R-DATA/SMDPP Transport
TIA/EIA-136-620	Teleservice Segmentation and Reassembly
TIA/EIA-136-630	Broadcast Teleservice Transport
TIA/EIA-136-7XX	Teleservices
TIA/EIA-136-700	Introduction to Teleservices
TIA/EIA-136-710	Short Message Service—Cellular Messaging Teleservice
TIA/EIA-136-720	Over-the-Air Activation Teleservice
TIA/EIA-136-730	Over-the-Air Programming Teleservice
TIA/EIA-136-750	General UDP Transport Service
TIA/EIA-136-9XX	Annexes/Appendices
TIA/EIA-136-900	Introduction
TIA/EIA-136-905	Normative Information
TIA/EIA-136-910	Informative Information

Layer 1: The Digital PCS Physical Layer

TIA/EIA-136 SPECIFIES the physical characteristics of the Digital PCS air interface. These physical characteristics include RF-related parameters, modulation format, power output requirements, TDMA frame and time-slot structures, and the channel coding and interleaving schemes used on the various Digital PCS channels. More than anything else, the physical layer sets Digital PCS apart from other air interface standards. The physical layer, also called layer 1, defines the hardware requirements of Digital PCS mobiles and base stations. It also influences many of the operational parameters of Digital PCS systems. These include the distance between cell sites, the number of RF channels available per cell or sector, the required transmit power at the mobile and base station, the cell site antenna height and gain, and the frequency reuse pattern.

This chapter describes the Digital PCS physical layer. It does not describe the physical layer of the analog control channel (ACC) or analog voice channel (AVC), as these physical characteristics are the same as AMPS and are well described elsewhere. The physical layers of both the digital control channel (DCCH) and the digital traffic channel (DTC) are described in [1]. The time-slot structure, channel coding, and interleaving differ for these two digital channel types, but the remaining physical layer characteristics are the same. This chapter does not describe the physical layer of 136+, as this is covered in a subsequent chapter.

4.1 Channelization and duplexing

TIA/EIA-136 specifies operation of Digital PCS in the 800-MHz and 1,900-MHz frequency bands. To distinguish these frequency bands from the individual allocations within each band, they are called *hyperbands*. The 800-MHz hyperband extends from approximately 824 MHz to approximately 894 MHz in the traditional North American cellular spectrum. There is a 45-MHz separation between mobile and base station transmissions in the 800-MHz hyperband. The 1,900-MHz hyperband extends from approximately 1,850 MHz to approximately 1,990 MHz in the worldwide PCS spectrum allocation. There is an approximate 80-MHz separation between mobile and base station transmissions in the 1,900-MHz hyperband. There is the potential for additional hyperbands to be defined for Digital PCS operation in the future, should the need arise.

Each hyperband is divided into RF channels of 30-kHz bandwidth. The RF channels are numbered sequentially within each hyperband. Duplex channel pairs are numbered the same—that is, the mobile transmit channel and corresponding base station transmit channel have the same channel number. There are 833 duplex channel pairs in the 800-MHz hyperband, and 1,999 duplex channel pairs in the 1,900-MHz hyperband. The first channel in both hyperbands and the last channel in the 1,900-MHz hyperband are not used because they serve as guard bands. Table 4.1 contains equations for calculating the center frequency of each 30-kHz RF channel from the channel number for both hyperbands.

Table 4.1
Calculation of Center Frequencies From Channel Numbers

Transmitter	Channel Number (C)	Center Frequency (MHz)
800-MHz mobile	$1 \leq C \leq 799$	0.030 C + 825.00
	$990 \leq C \leq 1{,}023$	0.030 (C − 1,023) + 825.00
800-MHz base station	$1 \leq C \leq 799$	0.030 C + 870.00
	$990 \leq C \leq 1{,}023$	0.030 (C − 1,023) + 870.00
1,900-MHz mobile	$1 \leq C \leq 1{,}999$	0.030 C + 1,849.980
1,900-MHz base station	$1 \leq C \leq 1{,}999$	0.030 C + 1,930.020

While TIA/EIA-136 allows for flexibility in the allocation of spectrum within the hyperbands, bands that are currently in use within the hyperbands have been identified in the standard. Two 800-MHz bands are identified, designated system A and system B. Both 800-MHz bands are allocated 416 duplex channels, corresponding to approximately 12.5 MHz each. Six 1,900-MHz bands are identified, designated bands A through F. The 1,900-MHz A, B, and C bands are allocated 15 MHz, while the D, E, and F bands are allocated 5 MHz. Table 4.2 summarizes the channelization and duplexing in the 800-MHz hyperband, and Table 4.3 does the same for the 1,900-MHz hyperband.

Table 4.2
Channelization and Duplexing in the 800-MHz Hyperband

System	Number of Channels	Channel Number Range	Mobile Transmit Center Frequency Range (MHz)	Base Station Transmit Center Frequency Range (MHz)
Not used	1	990	824.010	869.010
A″	33	991–1023	824.040–825.000	869.040–870.000
A	333	1–333	825.030–834.990	870.030–879.990
B	333	334–666	835.020–844.980	880.020–889.980
A′	50	667–716	845.010–846.480	890.010–891.480
B′	83	717–799	846.510–848.970	891.510–893.970

Table 4.3
Channelization and Duplexing in the 1,900-MHz Hyperband

Band	Number of Channels	Channel Number Range	Mobile Transmit Center Frequency Range (MHz)	Base Station Transmit Center Frequency Range (MHz)
Not used	1	1	1,850.010	1,930.050
A	497	2–498	1,850.040–1,864.920	1,930.080–1,944.960
A and D	3	499–501	1,864.950–1,865.010	1,944.990–1,945.050
D	164	502–665	1,865.040–1,869.930	1,945.080–1,949.970
D and B	2	666–667	1,869.960–1,869.990	1,950.000–1,950.030
B	498	668–1,165	1,870.020–1,884.930	1,950.060–1,964.970
B and E	2	1,166–1,167	1,884.960–1,884.990	1,965.000–1,965.030
E	165	1,168–1,332	1,885.020–1,889.940	1,965.060–1,969.980
E and F	2	1,333–1,334	1,889.970–1,890.000	1,970.010–1,970.040
F	164	1,335–1,498	1,890.030–1,894.920	1,970.070–1,974.960
F and C	3	1,499–1,501	1,894.950–1,895.010	1,974.990–1,975.050
C	497	1,502–1,998	1,895.040–1,909.920	1,975.080–1,989.960
Not used	1	1,999	1,909.950	1,989.990

4.2 Modulation format

The modulation format of the TIA/EIA-136 digital control channel and digital traffic channel is π/4 shifted, differentially encoded quadrature phase shift keying (π/4 DQPSK). Figure 4.1 shows the functional block diagram of a DQPSK modulator. Binary data is first converted to a parallel stream, then differentially encoded into two bit symbols with a symbol rate of 24.3 ksymbols/s. The resulting two encoded data sequences are passed through square-root raised cosine filters for pulse shaping, then used to quadrature modulate a carrier.

The differential encoding of the data is actually carried out as differential phase encoding according to the rules of Table 4.4. This allows for differential detection of the data at the receiver and limits the phase changes to minimize the linearity requirements of power amplifiers in the transmit chain. The differential phase encoder maps the X and Y data sequences to in-phase (I) and quadrature (Q) components at time k as follows

$$I_k = I_{k-1} \cos(\Delta\theta) - Q_{k-1} \sin(\Delta\theta) \qquad (4.1)$$

$$Q_k = I_{k-1} \sin(\Delta\theta) + Q_{k+1} \cos(\Delta\theta) \qquad (4.2)$$

where I_{k-1} and Q_{k-1} are the amplitudes at the previous pulse time (for the previous symbol).

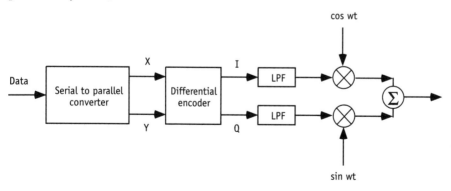

Figure 4.1 Functional block diagram of a DQPSK modulator.

Table 4.4
Phase Change Mapping in DQPSK Modulation

X	Y	Δθ
1	1	$-3\pi/4$
0	1	$3\pi/4$
0	0	$\pi/4$
1	0	$-\pi/4$

The square-root raised cosine filters in the in-phase and quadrature channels are low-pass filters with linear phase and frequency response given by

$$|H(f)| = \begin{cases} 1 & 0 \leq f \leq \dfrac{(1-\alpha)}{2T} \\ \sqrt{\dfrac{1}{2}\left\{1 - \sin\left[\dfrac{\pi(2fT-1)}{2\alpha}\right]\right\}} & \dfrac{(1-\alpha)}{2T} \leq f \leq \dfrac{(1+\alpha)}{2T} \\ 0 & f > \dfrac{(1+\alpha)}{2T} \end{cases} \quad (4.3)$$

where T is the symbol period (41.2 s) and α is the roll-off factor (0.35), as shown in Figure 4.2.

Ignoring for the moment the effects of the square-root raised cosine filters, it can be seen from the block diagram in Figure 4.1 that the transmitted signal across the duration of the current symbol is

$$S_k(t) = I_k \cos(w_c t) - Q_k \sin(w_c t) \quad (4.4)$$

Substituting (4.1) and (4.2) into (4.4) and simplifying through trigonometric identities gives the following relation:

$$S_k(t) = I_{k-1} \cos(w_c t + \Delta\theta) - Q_{k-1} \sin(w_c t + \Delta\theta) \quad (4.5)$$

Layer 1: The Digital PCS Physical Layer 53

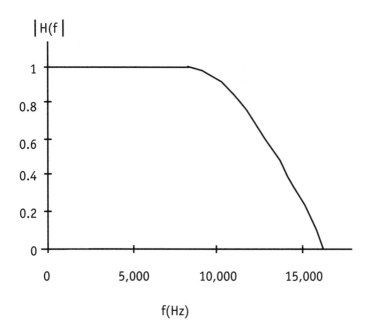

Figure 4.2 Frequency response of a square-root raised cosine filter.

Euler's theorem can be applied to (4.5) to yield

$$S_k(t) = S_{k-1}(t) e^{j\Delta\theta} \qquad (4.6)$$

Equation (4.6) is important because it shows the relationship between the signal transmitted in the current symbol interval and that transmitted in the previous symbol interval. All of the information is transmitted as a change in phase from the previously transmitted signal, and the phase change is limited to $\pm\pi/4$ or $\pm 3\pi/4$. This yields the phase constellation shown in Figure 4.3. If the transmitted signal is at one of the four states denoted by * in the current symbol interval, then for the next symbol interval the transmitted signal changes to one of the four states denoted by +. The fact that all of the information is transmitted as a difference in phase from the previous symbol allows for differential demodulation at the receiver, which has the advantage of hardware simplicity. A coherent demodulator could be used at the receiver, even

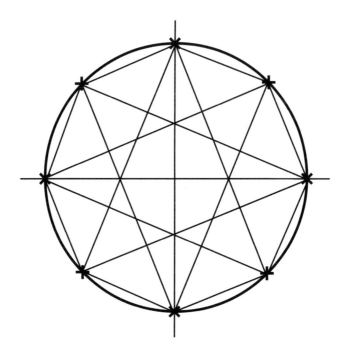

Figure 4.3 Phase constellation for DQPSK modulation.

though the data is differentially encoded, to improve receiver performance under low Doppler spread conditions [2].

4.3 Power output

The power output requirement of a Digital PCS mobile depends upon the *mobile station power class* of the mobile and the hyperband in which the mobile is operating. The TIA/EIA-136 standard defines the power output of the mobile in terms of the nominal *effective radiated power* (ERP) with respect to a half-wave dipole antenna, which is the average power transmitted by the mobile while in digital mode. There are four power classes for mobiles operating in the 800-MHz hyperband and two power classes for mobiles operating in the 1,900-MHz hyperband. The power output requirements for Digital PCS mobiles are summarized in Table 4.5.

Table 4.5
Summary of Digital PCS Mobile Power Output Requirements

Mobile Power Level	Nominal ERP (dBm) for the 800-MHz Hyperband				Nominal ERP (dBm) for the 1,900-MHz Hyperband	
	I	II	III	IV	II	IV
0	36	32	28	28	30	28
1	32	32	28	28	30	28
2	28	28	28	28	28	28
3	24	24	24	24	24	24
4	20	20	20	20	20	20
5	16	16	16	16	16	16
6	12	12	12	12	12	12
7	8	8	8	8	8	8
8	8	8	8	4 ± 3 dB	4 ± 3 dB	4 ± 3 dB
9	8	8	8	0 ± 6 dB	0 ± 6 dB	0 ± 6 dB
10	8	8	8	−4 ± 9 dB	−4 ± 9 dB	−4 ± 9 dB

Table 4.5 shows the ten mobile power levels available for Digital PCS mobiles. The TIA/EIA-136 standard mandates that under normal operating conditions a mobile must maintain its power level within the range of +2/−4 dB of nominal for power levels 0 to 7, and within +2/−6 dB of its initial level for power levels 8, 9, and 10. The mobile must be capable of changing its transmit power in response to a commanded change in power level of 4 dB within 20 ms. The base station identifies to the mobile the appropriate operating power level to use based on the mobile station power class. Power levels 8, 9, and 10 are only used in digital mode. Microcell and picocell operations exploit these lower power levels. The majority of Digital PCS mobiles are class IV mobiles, with 28 dBm maximum output power. RF planning for macrocells and many microcells is based on the use of class IV mobiles operating at their maximum output power on cell boundaries. The base station typically directs mobiles operating on digital traffic channels to reduce their power for closer in operation. In this manner, the cochannel interference level is minimized and the dynamic range requirements of the base station may be reduced.

Many of the power output requirements of Digital PCS base stations are not standardized. The maximum base station output power is typically governed by national regulatory agencies. The base station output power also varies depending upon the type of operation planned for the base station. For example, a base station supporting a microcell does not need to operate at as high an output power as a base station supporting a macrocell. The TIA/EIA-136 standard does mandate that the output power of the base station must be maintained within +1/−3 dB of the nominal level. Prior to 136+, it was also mandated that if a base station is transmitting on one or more time slots of an RF channel, it must transmit constant power on all of the time slots on that channel. This ensured that mobiles were able to use any and all of the time slots transmitted by the base station to maintain synchronization, train an equalizer, and perform automatic gain control. The constant power requirement of base stations was removed with 136+ to facilitate *downlink power control*, which allows the base station to transmit at different power levels on each time slot of an RF channel. Downlink power control can reduce the interference level caused by base station transmissions and lead to tighter frequency reuse. Without the constant power requirement, Digital PCS mobiles must only rely upon information available within their assigned time slot(s) to perform the aforementioned functions.

IS-137 and IS-138 define the emissions masks for Digital PCS mobiles and base stations, respectively [3,4]. The emissions mask guarantees that the transmitted power is confined as much as possible to the desired channel. This reduces the potential for interference to users operating on adjacent and alternate channels. An adjacent channel is either of the channels centered 30 kHz from the center frequency. An alternate channel is either of the channels centered 60 kHz from the center frequency. The second alternate channel is either of the channels centered 90 kHz from the center frequency. The emissions masks for Digital PCS mobiles and base stations are shown in Figure 4.4.

4.4 Frame and time-slot structures

The frame structure is the primary physical layer characteristic of Digital PCS that allows for TDMA operation. A single TDMA frame is 40 ms in

Layer 1: The Digital PCS Physical Layer 57

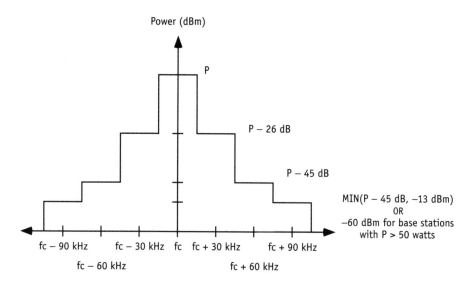

Figure 4.4 Emissions masks for Digital PCS mobiles and base stations.

length and is composed of six time slots, labeled 1 through 6, as depicted in Figure 4.5. A full-rate TDMA channel is composed of a time-slot pair separated by 20 ms. For example, time slot 1 is paired with time slot 4 to form one full-rate TDMA channel. Time slot 2 is paired with time slot 5, and time slot 3 is paired with time slot 6, to form additional full-rate TDMA channels. None of the time slots in a TDMA frame are paired in half-rate operation. A single RF channel can also support double full-rate operation occupying four of the six time slots in a TDMA frame. All six time slots are allocated to one user in triple full-rate operation. There has to be a compelling reason to devote a full RF channel to a single user, though, as this somewhat defeats the purpose of TDMA.

Much flexibility exists in the types of channels that may be allocated to a TDMA frame. For example, time slots 1 and 4 may contain a full-rate digital control channel while time slots 2 and 5 and time slots 3 and 6 may contain two digital traffic channels. If a DCCH is present on an RF channel, it must begin in time slot 1. Only half-rate and full-rate DCCHs are defined for TIA/EIA-136, although more than one DCCH may be allocated to an RF channel. Typical Digital PCS implementations allow for one RF channel per sector containing a full-rate DCCH and two full-rate

58 Understanding Digital PCS

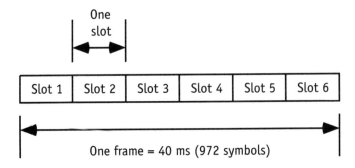

Figure 4.5 TDMA frame structure.

DTCs, with the remaining digital channels in the sector consisting of three full-rate DTCs.

An offset between the reverse and forward channel frame timing exists so that a mobile is not required to transmit and receive at the same instant of time while operating on a full-rate or half-rate channel. Because of this offset, a duplexer is not needed in the mobile unless AMPS is supported or double or triple full-rate operation is required. This is particularly important for mobiles operating in the 1,900-MHz hyperband where AMPS is not supported. As shown in Figure 4.6, the offset between the reverse and forward frame timing is one time slot plus 45 symbols.

The offset between the reverse and forward TDMA frames may be changed through a process called *time alignment* on the digital traffic channel. Time alignment allows for the advancing of the offset between the reverse and forward frames by up to 15 symbols, in half-symbol units. The reason for this advance is to allow a mobile operating far from the base station to transmit in a time slot without interfering with a mobile operating close to the base station and transmitting in the next time slot. In this case, the timing advance is applied to the mobile operating far from the base station. A 15-symbol timing advance allows for a maximum cell radius approximately 185 km, as calculated below:

$$(15 \text{ symbols}) \times (41.2 \times 10^{-6} \text{ s/symbol}) \times (3 \times 10^{5} \text{ km/s}) = 185.4 \text{ km}$$

Layer 1: The Digital PCS Physical Layer   59

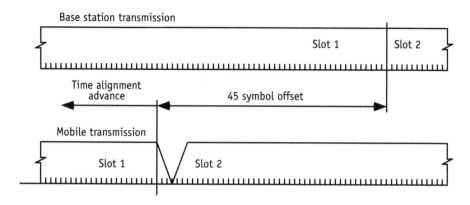

Figure 4.6 Timing offset between reverse and forward channel frames.

4.4.1 Digital control channel time-slot structure

Two timing definitions, the *superframe* and the *hyperframe*, are important for the operation of the forward digital control channel. A superframe consists of 16 TDMA frames, and a hyperframe consists of two superframes. This means that a superframe is 640 ms in duration, and a hyperframe is 1.28 seconds in duration. The relationship between TDMA frames, superframes, and hyperframes is depicted in Figure 4.7. For a full-rate DCCH, there are 32 time slots in a superframe because two time slots per TDMA frame are used for the DCCH. The structure of the DCCH is based on the superframe and hyperframe timings. The implementation of sleep mode, as described in Chapter 10, also relies on the superframe and hyperframe structure of the DCCH.

Different time-slot structures are used on the forward and reverse digital control channel. In the reverse direction, from the mobile to the base station, there is a normal slot format and an abbreviated slot format. The abbreviated slot format is used for large cell sites to ensure that mobiles transmitting on adjacent time slots do not overlap each other when one mobile is located at the cell boundary and the other mobile is located close to the base station. In the forward direction, from the base station to the mobile, there is only one time-slot format. These time-slot formats are shown in Figure 4.8. The time-slot fields are defined in Table 4.6.

60 Understanding Digital PCS

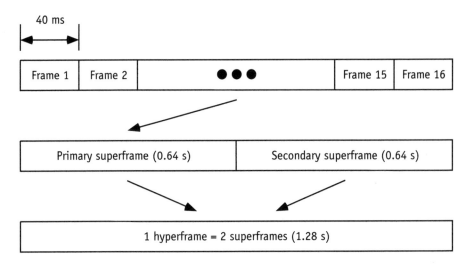

Figure 4.7 Superframe and hyperframe structure on the forward DCCH.

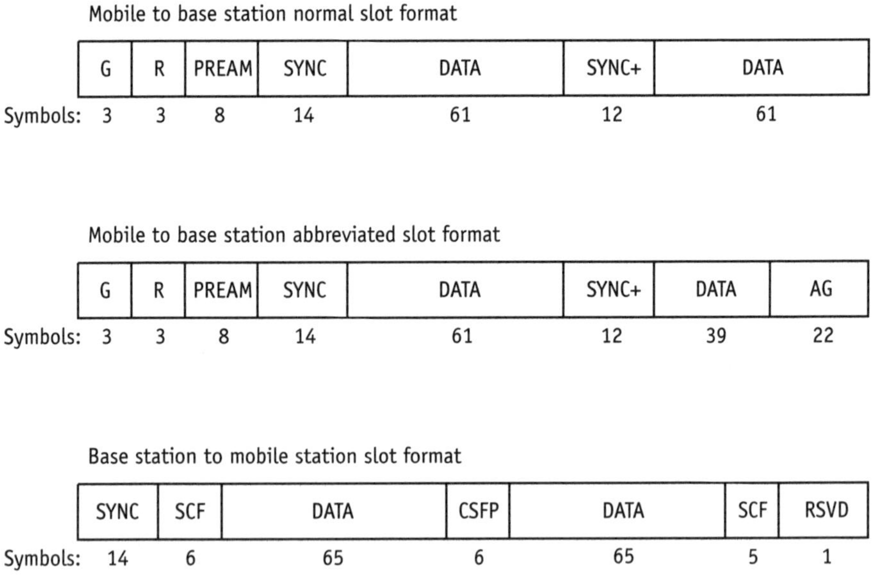

Mobile to base station normal slot format

G	R	PREAM	SYNC	DATA	SYNC+	DATA
3	3	8	14	61	12	61

Symbols:

Mobile to base station abbreviated slot format

G	R	PREAM	SYNC	DATA	SYNC+	DATA	AG
3	3	8	14	61	12	39	22

Symbols:

Base station to mobile station slot format

SYNC	SCF	DATA	CSFP	DATA	SCF	RSVD
14	6	65	6	65	5	1

Symbols:

Figure 4.8 DCCH time-slot formats.

Table 4.6
Summary of DCCH Time-Slot Fields

DCCH Field	Interpretation	Definition
AG	Abbreviated guard	Extra guard time in the abbreviated slot format on the reverse DCCH, used to ensure mobiles transmitting at the cell border do not interfere with close in mobiles transmitting on adjacent time slots.
CSFP	Coded superframe phase	Coded identification of the time slot in relation to the start of a superframe, used by mobiles to find the start of the superframe.
DATA	Coded information bits	Coded logical channel information.
G	Guard	Guard time to reduce the probability that mobiles transmitting on adjacent time slots interfere with each other.
PREAM	Preamble	Preamble used by the base station to perform automatic gain control and facilitate symbol synchronization. The eight symbols in the field each have $-\pi/4$ phase changes.
R	Ramp time	Power-up or power-down ramp intervals, used to reduce spurious transmissions.
RSVD	Reserved	Unused bits.
SCF	Shared channel feedback	Feedback used to control mobile access on the reverse DCCH with three subfields: Busy/Reserved/Idle (BRI), Coded Partial Echo (CPE), and Received/Not Received (R/N).
SYNC	Synchronization	Synchronization field used for slot synchronization, equalizer training, and time-slot identification. The same synchronization sequences available for use on the DTC are used on the DCCH.
SYNC+	Additional synchronization	Additional synchronization information to improve base station receiver performance. The field is specified by the following sequence of phase changes: $\pi/4, -\pi/4, 3\pi/4, -3\pi/4, -\pi/4, -\pi/4, -3\pi/4, 3\pi/4, 3\pi/4, \pi/4, \pi/4, -\pi/4$.

Logical channels are specified for the forward and reverse DCCH, as shown in Figure 4.9. Only one logical channel is defined for the reverse DCCH—the random access control channel (RACH, pronounced to rhyme with *scratch*). The RACH is a shared, point-to-point, acknowledged logical channel used by mobiles to access the DCCH. Collision avoidance and contention resolution on the RACH is provided through the SCF field on the forward DCCH. A mobile reads the BRI subfield to determine when to begin an access attempt. After the first burst on the RACH, the mobile reads the PE subfield to determine if its access was captured. If a burst has been transmitted by the mobile on the RACH, the mobile reads the R/N subfield to determine if the burst has been received.

The logical channels on the forward DCCH are the SMS point-to-point, paging, and access response channel (SPACH, also pronounced to rhyme with *scratch*); broadcast control channel (BCCH); SCF as described above; and a reserved channel defined for future use. The SPACH is further divided into the SMS channel (SMSCH), paging channel (PCH), and access response channel (ARCH). Likewise, the BCCH is further divided into the fast broadcast control channel (F-BCCH), extended broadcast control channel (E-BCCH), and SMS broadcast

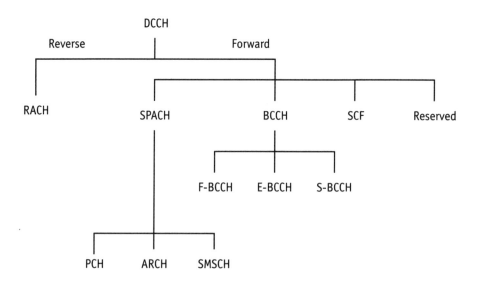

Figure 4.9 DCCH logical channels.

control channel (S-BCCH). The SPACH is a shared, point-to-point channel. The ARCH and the SMSCH may be used in acknowledged mode through the use of automatic retransmission request (ARQ), described in Chapter 5. The BCCH is a shared, point-to-multipoint, unacknowledged channel.

The SMSCH delivers teleservice-related messages to mobiles, the PCH delivers pages and orders to mobiles, and the ARCH delivers responses to messages from the mobile. The F-BCCH carries time-critical information essential for mobiles to access the network, while the E-BCCH carries less time-critical information about the network. The S-BCCH carries broadcast teleservices. A fixed number of time slots per superframe on the forward DCCH are allocated for use as F-BCCH, E-BCCH, S-BCCH, reserved channels, and SPACH. The F-BCCH is repeated every superframe, while the E-BCCH may cycle across many superframes. Some SPACH slots may be restricted from allocation as PCH subchannels. The minimum and maximum slot allocations per superframe for the forward DCCH are summarized in Table 4.7, and the superframe and hyperframe structure in terms of these logical channels is shown in Figure 4.10.

4.4.2 Digital traffic channel time-slot structure

Different time-slot structures are also used on the forward and reverse digital traffic channel, as shown in Figure 4.11. The time-slot fields are defined in Table 4.8. The DCCH locator portion of the CDL identifies a range of 8 channels in the 800-MHz hyperband and 16 channels in the

Table 4.7
Full-Rate DCCH Slot Allocations per Superframe

Logical Channel	Minimum Slots	Maximum Slots
F-BCCH (F)	3	10
E-BCCH (E)	1	8
S-BCCH (S)	0	15
Reserved (R)	0	7
SPACH	2	28

64 Understanding Digital PCS

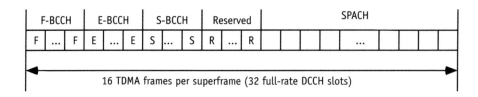

Figure 4.10 Superframe and hyperframe structure on the forward DCCH (full-rate).

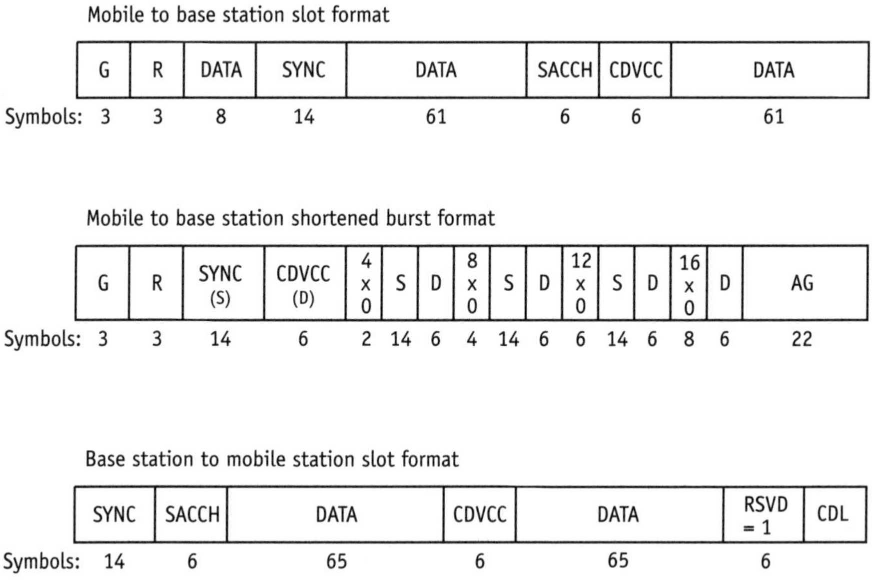

Mobile to base station slot format

G	R	DATA	SYNC	DATA	SACCH	CDVCC	DATA
3	3	8	14	61	6	6	61

Symbols: 3 3 8 14 61 6 6 61

Mobile to base station shortened burst format

G	R	SYNC (S)	CDVCC (D)	4x0	S	D	8x0	S	D	12x0	S	D	16x0	D	AG
3	3	14	6	2	14	6	4	14	6	6	14	6	8	6	22

Base station to mobile station slot format

SYNC	SACCH	DATA	CDVCC	DATA	RSVD = 1	CDL
14	6	65	6	65	6	

Figure 4.11 DTC time-slot formats.

1,900-MHz hyperband where a DCCH may be found. A mobile scans RF channels for a DCCH by use of the DCCH locator to more rapidly find a DCCH if the mobile encounters a DTC in the search. The digital verification color code portion of the CDVCC can take on one of 255 values to uniquely identify the base station from which the mobile is obtaining service.

Table 4.8
Summary of DTC Time-Slot Fields

DCCH Field	Interpretation	Definition
CDL	Coded DCCH locator	Coded field indicating a range of channels in which a DCCH may be found
CDVCC	Coded digital verification color code	Coded field containing a digital verification color code used to indicate that the correct rather than cochannel data is being decoded
DATA	Coded information bits	Coded voice, user data, or fast associated control channel
G	Guard	Guard time to reduce the probability that mobiles transmitting on adjacent time slots interfere with each other
R	Ramp time	Power-up or power-down ramp intervals, used to reduce spurious transmissions
RSVD	Reserved	Unused bits
SACCH	Slow associated control channel	A continuous channel used for signaling message exchange
SYNC	Synchronization	Synchronization field used for slot synchronization, equalizer training, and time-slot identification

The shortened burst on the DTC is similar to the abbreviated burst on the DCCH. A base station typically commands a mobile to transmit using the shortened burst format if the mobile does not have proper time alignment information on initial channel assignment or handoff. The special pattern of the shortened burst format helps the base station to quickly determine the offset of the mobile's burst from the ideal time relationship and send time alignment information to the mobile. Since no user data is transmitted in a shortened burst, it behooves the base station to send time alignment information to the mobile as rapidly as possible so the mobile can begin transmitting using the normal burst format in which voice or user data is sent.

The DATA field may contain voice, user data, or the fast associated control channel (FACCH). The FACCH is a blank-and-burst channel used for signaling exchange between the mobile and base station. Blank-and-burst means that the signaling preempts voice or user data in the time slot. The SACCH is used instead of the FACCH for signaling requiring low data rates and is attractive because it does not preempt voice or user data. Most messages on the DTC have the option of being carried on either the SACCH or FACCH. These control channels are called *associated* because they have a one-to-one correspondence with the DTC, as opposed to ACCs and DCCHs, which do not.

Six unique synchronization sequences are defined for the SYNC field, as shown in Table 4.9. Each sequence has good autocorrelation properties to facilitate synchronization and equalizer training. The preferred assignment of synchronization sequences on an RF channel is one per DTC user. For example, if a channel is serving three full-rate DTC users, each DTC would be allocated a different synchronization sequence, starting with sync 1. Assigning different sequences to different DTCs on the same RF channel helps the mobile and base station to identify the correct DTC. For this reason, the synchronization sequence is also called the time-slot identifier.

4.5 Channel coding and interleaving

The most important fields on the DCCH and DTC are protected against the adverse effects of multipath fading, interference, and other RF channel impairments through the use of channel coding. Channel coding adds redundancy to the transmission by forming code words so the receiver is more likely to properly decode the information. Interleaving is also used in the DATA field to minimize the effects of burst errors. Interleaving spreads the information bits across a time slot or multiple time slots so that if a burst of errors occurs in one part of the slot it will not corrupt a large number of contiguous information bits. This makes the channel coding more useful, because there are fewer errors per code word to detect and correct.

Some of the fields on the DCCH and DTC are protected by what is called a Hamming code. A Hamming code is a parity-check code that can

Table 4.9
Definition of Synchronization Sequences in Terms of Phase Changes

Sync	Phase Changes													
1	$-\pi/4$	$-\pi/4$	$-3\pi/4$	$\pi/4$	$3\pi/4$	$-3\pi/4$	$-3\pi/4$	$-\pi/4$	$3\pi/4$	$\pi/4$	$-\pi/4$
2	$-3\pi/4$	$-3\pi/4$	$-3\pi/4$	$-3\pi/4$	$-3\pi/4$	$-3\pi/4$	$-3\pi/4$	$-3\pi/4$	$-3\pi/4$	$-3\pi/4$	$-3\pi/4$
3	$-3\pi/4$	$-3\pi/4$	$-3\pi/4$	$-3\pi/4$	$-3\pi/4$	$-3\pi/4$	$-3\pi/4$	$-3\pi/4$	$-3\pi/4$	$-3\pi/4$	$-3\pi/4$
4	$-3\pi/4$	$-3\pi/4$	$-3\pi/4$	$-3\pi/4$	$-3\pi/4$	$-3\pi/4$	$-3\pi/4$	$-3\pi/4$	$-3\pi/4$	$-3\pi/4$	$-3\pi/4$
5	$-3\pi/4$	$-3\pi/4$	$-3\pi/4$	$-3\pi/4$	$-3\pi/4$	$-3\pi/4$	$-3\pi/4$	$-3\pi/4$	$-3\pi/4$	$-3\pi/4$	$-3\pi/4$
6	$-3\pi/4$	$-3\pi/4$	$-3\pi/4$	$-3\pi/4$	$-3\pi/4$	$-3\pi/4$	$-3\pi/4$	$-3\pi/4$	$-3\pi/4$	$-3\pi/4$	$-3\pi/4$

detect and correct single errors in a block of information bits. Other fields are protected by a cyclic redundancy check (CRC) code. A CRC code is another parity-check code capable of detecting multiple errors in a block of information bits. For both these codes, similar algorithms can be used to generate the parity bits that are added to the information bits to form the code words, and similar techniques can be used to decode the information.

The DATA fields on the DCCH and the DTC are protected by a *convolutional code*. The Hamming and CRC codes are examples of block codes for which parity bits are calculated based on a block of information bits. Conversely, a convolutional encoder operates on a span of information bits that is shifted one information bit every time another information bit is input to the encoder. The length of the span in bits is known as the constraint length. The larger the constraint length, the better the robustness of the code to burst errors. The rate of the convolution code is the ratio of the number of information bits input to the encoder to the number of coded bits output from the encoder, and is always less than one. The lower the rate, the more redundancy is added to the information bits and the more error protection is afforded.

Interleaving of the information bits occurs after convolutional coding. Interleaving is typically accomplished by entering the coded bits into an array row-wise (or column-wise) and reading them out column-wise (or row-wise).

4.5.1 Digital control channel

The CSFP, CPE, and DATA fields on the DCCH are encoded. The superframe phase is 8 bits long (the most significant three bits are set to zero) and Hamming encoding is used to calculate 4 parity bits. The 12-bit CSFP code word is composed of the 8 bits of superframe phase and the 4 inverted parity bits. The parity bits are inverted to help mobiles discriminate between the CSFP and the CDVCC. The CPE is encoded in a similar manner, but with the most significant bit of the partial echo removed to make the CPE field 11 bits in length. Rate 1/2 convolutional coding is used in the DATA field on the DCCH.

Coded bits are placed in the DATA field after interleaving them across one time slot. The interleaving is accomplished by grouping the coded bits into 20 segments of 13 bits in length. The segments are placed column-wise into an array, then read row-wise from the array and placed into the DATA field.

4.5.2 Digital traffic channel

The CDL, CDVCC, DATA, SACCH, and FACCH fields on the DTC are encoded. The CDL is encoded exactly as the CPE on the DCCH. The CDVCC is encoded exactly as the CSPF on the DCCH, except that the parity bits are not inverted. Rate 1/2 convolutional coding is used for SACCH transmissions. The SACCH data is interleaved over 12 consecutive DTC slots to maximize its robustness. Rate 1/4 convolutional coding is used for FACCH transmissions, and the FACCH data is interleaved across two time slots. The channel coding and interleaving of the DATA field differ based on the application. For example, the channel coding for voice and data services on the DTC is covered in Chapter 11.

References

[1] Telecommunications Industry Association, *TIA/EIA Interim Standard: TDMA Cellular/PCS—Radio Interface—Mobile Station-Base Station Compatibility, Revision A, TIA/EIA/IS-136-A*, Oct. 1996.

[2] Liu, C., and K. Feher, "π/4-QPSK Modems for Satellite Sound/Data Broadcast Systems," *IEEE Transactions on Broadcasting*, Vol. 37, No. 1, Mar. 1991, pp. 1–8.

[3] Telecommunications Industry Association, *TIA/EIA Interim Standard: TDMA Cellular/PCS—Radio Interface—Minimum Performance Standard for Mobile Stations, Revision A, TIA/EIA/IS-137-A*, 1996.

[4] Telecommunications Industry Association, *TIA/EIA Interim Standard: TDMA Cellular/PCS—Radio Interface—Minimum Performance Standards for Base Stations, Revision A, TIA/EIA/IS-138-A*, 1996.

Layer 2: The Digital PCS Data Link Layer

5.1 The OSI reference model

Layer 2 of Digital PCS is the data link layer, as defined in the seven-layer Open Systems Interconnection (OSI) reference model [1]. The OSI reference model was developed by the International Organization for Standardization (ISO) to provide a framework for the development of standards for interconnecting two or more systems. A basic understanding of the seven-layer OSI reference model, shown in Figure 5.1, will help in interpreting this chapter. The model defines a layered architecture wherein each *layer* performs a different function and depends upon the layer below it to provide services. Services are made available to the higher layer by *service primitives* provided at specific *service access points*. Communication between the same layers in two systems is provided by a *protocol*.

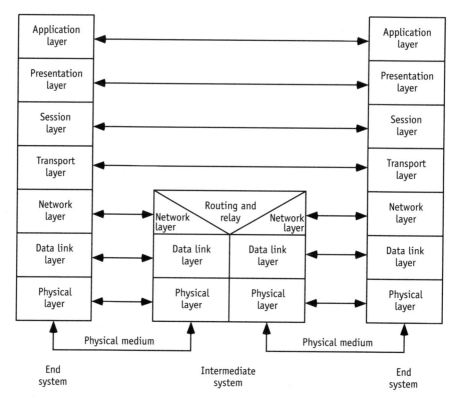

Figure 5.1 The seven-layer OSI reference model.

The layers of the OSI reference model are often described in terms of data communications between two *end systems* via an *intermediate system* that performs routing and relay functions. The application layer provides the interface to the user and may be thought of as the application programs residing on the end systems. The presentation layer defines the syntax of all data exchanged between the application layers of the end systems. Organization and synchronization of the data transferred between the presentation layers of the end systems are provided by the session layer. The transport layer optimizes the use of network layer services and ensures the quality of the data transferred between the session layers of the end systems. The network layer provides a means to establish, maintain, and terminate connections between end systems. The purpose of the data link layer is to provide addressing; frame delimiting; error detection, recovery, and sequencing; media access control; and flow control of data transferred between network layer entities. Finally, the physical layer provides a mechanism to send and receive a stream of bits over a physical transmission medium. The layers are often numbered from one to seven, starting with the physical layer.

The TIA/EIA-136 standard defines all aspects of layer 1 (physical layer), layer 2 (data link layer), and layer 3 (network layer) for the Digital PCS air interface [2]. Higher layer Digital PCS functionality is provided through teleservices, as described in Chapter 12. As can be seen from Figure 5.1, intermediate systems do not communicate above layer 3, but only relay data. This is key to the teleservice concept. Chapters 4 and 6 describe layer 1 and layer 3, respectively, of the DCCH and the DTC. The layer 2 is well defined for the DCCH, but only minimally so for the DTC. This is because the DTC is used to transport both voice and data services, which have different layer 2 requirements. Furthermore, different data services may require different layer 3 functions. Layer 2 for asynchronous data and group 3 fax service on a DTC is described in Chapter 13.

5.2 Layer 2 of the DCCH

Layer 2 of the DCCH provides addressing; frame delimiting; error detection, recovery, and sequencing; media access control; and flow control on the forward and reverse DCCH, through the mechanisms shown in Table 5.1. Service access points, service primitives, and protocols are defined for the logical channels that comprise the forward and reverse DCCH. The service access points and service primitives constitute the window into layer 2 from layer 3. The protocols specify how a mobile and base station communicate with one another at layer 2.

5.2.1 DCCH layer 2 service access points and service primitives

Figure 5.2 shows the layer 2 service access points and service primitives for the DCCH. There are two layer 2 service access points at the mobile: a forward DCCH service access point (FDCCH SAP) and a reverse DCCH service access point (RDCCH SAP). At the base station, there are layer 2 service access points for each logical channel: fast broadcast control

Table 5.1
DCCH Layer 2 Functions

OSI Reference Model Data Link Layer Function	Functional Mapping to DCCH Layer 2
Addressing	Mobile station ID (MSID) management
Frame delimiting	Header formatting Layer 3 message concatenation Filler packet data units Zero fill Layer 2 frame segmentation and reassembly Burst usage
Error detection, recovery, and sequencing	CRC generation/verification Monitoring of radio link quality (MRLQ) Retransmission control (ARQ)
Media access control	Shared channel feedback Random and reserved access
Flow control	Shared channel feedback

channel service access point (F-BCCH SAP); extended broadcast control channel service access point (E-BCCH SAP); SMS broadcast channel service access point (S-BCCH SAP); SMS point-to-point, paging, and access response channel service access point (SPACH SAP); and RDCCH SAP.

To initiate a mobile access attempt on the RACH, a RDCCH Request primitive is sent from layer 3 to layer 2 within the mobile. The RDCCH Request primitive contains one or more layer 3 messages associated with the access attempt, an indication of the number of layer 3 messages and the length of each, and a message encryption indicator to identify the type of encryption to be applied to the layer 3 messages.

When a mobile correctly receives a layer 3 message on the FDCCH, the mobile sends an FDCCH Indication primitive from layer 2 to layer 3. The FDCCH Indication primitive contains the received layer 3 message, an indication of the length of the message, and an indication of the type of layer 3 message (for example, F-BCCH, E-BCCH, SMSCH, PCH, or ARCH). If the received layer 3 message is a SPACH message addressed to

Layer 2: The Digital PCS Data Link Layer 75

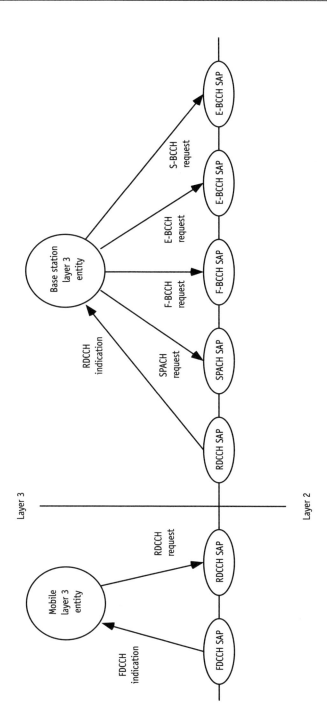

Figure 5.2 DCCH layer 2 service access points and primitives.

the mobile, then the FDCCH Indication primitive also includes an indication of the type of address in the form of a mobile station ID (MSID) that was used to send the message to the mobile.

When a base station receives a layer 3 message from a mobile, it sends an RDCCH Indication primitive from layer 2 to layer 3. The RDCCH Indication primitive includes one or more received layer 3 messages, an indication of the length of each message, the MSID from which the message came, and a message encryption indicator to identify the form of encryption used with the received messages.

To initiate an F-BCCH or E-BCCH message transmission, a base station sends an F-BCCH Request primitive or E-BCCH Request primitive from layer 3 to layer 2. The primitive includes the layer 3 message and an indication of the length of the message. To initiate a SPACH message transmission, a base station sends a SPACH Request primitive from layer 3 to layer 2. The SPACH Request primitive may or may not include a layer 3 message. If it does, it must also include an indication of the length of each message and a message encryption indicator. The primitive also includes up to five MSIDs, a Polling Indicator, and a burst usage indicator. The Polling Indicator is used for retransmission control and will be explained in greater detail later in this chapter. The burst usage indicates whether the SPACH Request primitive is intended for a PCH, ARCH, or SMSCH logical channel.

5.2.2 DCCH layer 2 protocols

The layer 2 protocol for a DCCH logical channel is specified by the allowable layer 2 frames and the rules that govern the use of the fields inside them. The general layer 2 frame structure is shown in Figure 5.3. A layer 2 frame fits inside a layer 1 DATA field after channel coding is applied. Because rate 1/2 channel coding is used on the DCCH, the length of a layer 2 frame is 1/2 the length of the layer 1 DATA field. This means that for a normal slot format on the reverse DCCH, a layer 2 frame is 122 bits in length. For an abbreviated slot format on the reverse DCCH, a layer 2 frame is 100 bits in length. All layer 2 frames on the forward DCCH are 130 bits in length. The last 5 bits of the layer 2 frame are always set to zero and used as tail bits into the channel encoder, so the actual layer 2 frame length is 117 bits for normal length bursts on the reverse DCCH, 95 bits

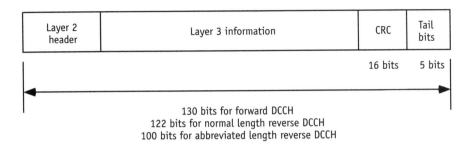

Figure 5.3 General DCCH layer 2 frame structure.

for abbreviated length bursts on the reverse DCCH, and 125 bits for the forward DCCH. The last 16 bits of these layer 2 frames are a CRC code for error detection.

The only logical channel on the reverse DCCH is the RACH. There are five types of RACH layer 2 frames, as shown in Figure 5.4: BEGIN and END, BEGIN, CONTINUE, END, and SPACH ARQ STATUS. The BEGIN and END frame is used when the mobile can fit one or more layer 3 messages destined for the base station into one layer 2 frame. The BEGIN frame is used by the mobile as the initial frame in a transaction requiring two or more frames to complete. The CONTINUE frame is used by the mobile as an intermediate frame in a transaction requiring three or more frames to complete. The END frame is used by the mobile as the last frame in a multiframe transaction. The SPACH ARQ STATUS frame is used by the mobile to report partial or complete status of an ARQ based transmission received by the mobile on the SPACH. The use of the SPACH ARQ STATUS frame will be explained more fully later in this chapter.

The Burst Type (BT) field indicates which one of the five types of RACH layer 2 frames is used in the frame. The Identity Type (IDT) field indicates which one of the four types of MSIDs is used in the frame. The types of MSIDs that a mobile may have are explained in Chapter 8. The Extension Header Indicator (EHI) field identifies whether or not an extension header is present in the frame. If the extension header is present, it immediately follows the EHI field. The extension header contains information used to identify the message encryption used to cover the layer 3 information. The Number of Layer 3 Messages (NL3M) field indicates the number of layer 3 messages included in the layer 2 frame. A

78 Understanding Digital PCS

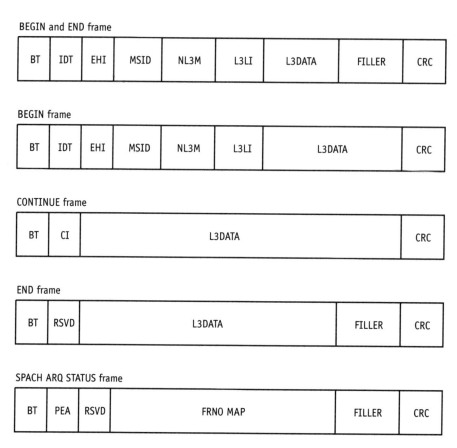

Figure 5.4 Five RACH layer 2 frame types.

Layer 3 Length Indication (L3LI) is included for each layer 3 message and defines the length of the message in octets. The Layer 3 Data (L3DATA) field contains the layer 3 messages. The Change Indicator (CI) field toggles for every new transmitted frame. The Partial Echo Assigned (PEA) field repeats the PEA value assigned to the mobile by the base during an ARQ mode transaction. The frame number map (FRNO MAP) provides a partial or complete bit map representation of the receive status of an ARQ mode transaction. Each frame is padded with filler to the end of the frame, if required.

The layer 2 frames for the F-BCCH and E-BCCH are similar. There are two basic types of F-BCCH and E-BCCH layer 2 frames, as shown in Figure 5.5: a BEGIN frame and a CONTINUE frame. The BEGIN frame

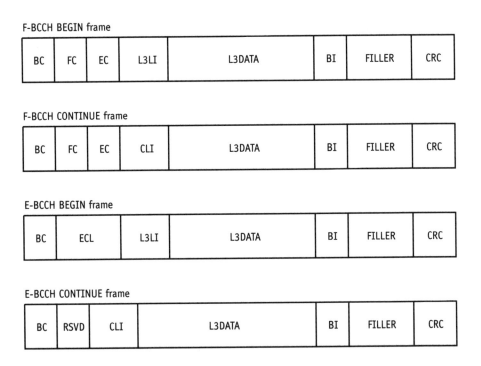

Figure 5.5 Two basic F-BCCH and E-BCCH layer 2 frame types.

is used by the base station to start the delivery of one or more layer 3 messages on the F-BCCH or E-BCCH. The CONTINUE frame is used for the continuation of a layer 3 message that is too long to fit into the previous frame. The fields within these two basic frame types can be varied to create other frame types. For example, a more than one layer 3 message can be carried within a BEGIN frame.

The Begin/Continue (BC) field indicates whether the frame is a BEGIN or CONTINUE frame. The F-BCCH Change (FC) and E-BCCH Change (EC) fields toggle to indicate a change to the F-BCCH and E-BCCH contents, respectively. The E-BCCH Cycle Length (ECL) field indicates the total number of layer 2 frames required for the current E-BCCH cycle. The Continuation Length Indicator (CLI) field indicates the number of bits in the current layer 2 frame used to carry information from a previously initiated layer 3 message. The Begin Indication (BI) field is used to indicate if a new layer 3 message starts in the same frame that another layer 3 message ends. The remainder of the fields in the

F-BCCH and E-BCCH frames are defined the same as those in the RACH frames.

There are many more types of SPACH layer 2 frames than for any of the other logical channels on the DCCH. This is because the SPACH is further subdivided into three logical channels used to carry different types of layer 3 messages. The possible frame formats for each of the SPACH logical channels are summarized in Table 5.2. In addition to the basic frame types listed in the table, a Null frame is defined for transmission on the SPACH when the base station has nothing else to send. Examples of these frame types are shown in Figure 5.6, along with the headers that may accompany them.

The SPACH Header A contains burst usage information and flags for managing mobile stations in sleep mode. The SPACH Header B contains supplementary header information used to identify the remaining contents of the layer 2 frame, as well as information about the layer 2 access mode to be used in the next access attempt made by the receiving mobile. The Extension Header contains supplementary header information used to identify the message encryption covering the layer 3 messages. Hard page frames are used to page mobiles on the PCH, and do not contain any layer 3 messages. Single MSID, double MSID, triple MSID, and

Table 5.2
Possible Layer 2 Frame Formats for SPACH Logical Channels

Frame Type	SMSCH	PCH	ARCH
Single MSID	Yes	Yes	Yes
Double MSID	No	Yes	Yes
Triple MSID	No	Yes	Yes
Quadruple MSID	No	Yes	Yes
Hard triple page	No	Yes	No
Hard quadruple page	No	Yes	No
Hard penta page	No	Yes	No
User group	No	Yes	Yes
CONTINUE	Yes	Yes	Yes
ARQ mode BEGIN	Yes	No	Yes
ARQ mode CONTINUE	Yes	No	Yes

Layer 2: The Digital PCS Data Link Layer   81

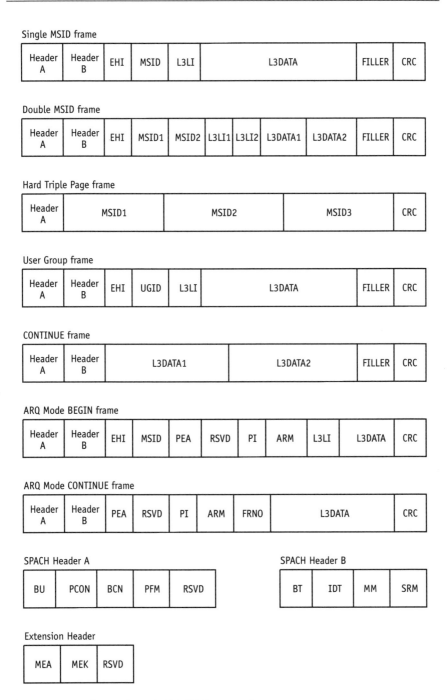

Figure 5.6   Examples of SPACH layer 2 frame types.

quadruple MSID frames are used to start the delivery of layer 3 messages in non-ARQ mode. The CONTINUE frame is used to continue layer 3 messages that are too long to fit into a single frame. The User Group frame is used to transmit a layer 3 message to a user group, which is a group of mobiles. The ARQ Mode BEGIN and ARQ Mode CONTINUE frames are used to send layer 3 messages to a mobile in ARQ mode.

The Burst Usage (BU) field identifies the SPACH layer 2 frame type. The PCH Continuation (PCON) field identifies if a mobile is required to continue reading time slots for a page. The BCCH Change Notification (BCN) field toggles every time there is a change to the F-BCCH or E-BCCH information. The User Group ID (UGID) field identifies the intended user group in a user group frame. The Message Mapping (MM) field indicates if the layer 3 messages in a frame are intended for one or more mobiles. The PEA field provides a partial echo value to a mobile for use during an ARQ mode transaction. The Polling Indicator (PI) field indicates whether or not the base station is soliciting a response (ARQ STATUS frame) from the mobile. The SPACH Response Mode (SRM) field indicates how a mobile is to respond once it has received all frames associated with a given SPACH message. The Message Encryption Algorithm (MEA) and Message Encryption Key (MEK) fields provide information on the message encryption used to cover layer 3 messages. The ARQ Response Mode (ARM) field indicates how a mobile is to respond once it has received an ARQ frame with the polling indicator set. The Frame Number (FRNO) field uniquely identifies specific frames of an ARQ mode transaction. The Go Away (GA) flag indicates whether the DCCH is barred. The remainder of the fields in the SPACH layer 2 frames are defined the same as those in the RACH, F-BCCH, and E-BCCH frames.

5.2.3 DCCH layer 2 media access control

Mobile access on the RACH may be contention- or reservation-based. Contention-based access is also known as random access. The base station controls all mobile access attempts through shared channel feedback and random access parameter settings at layer 3. A simplified flowchart of the mobile access process is shown in Figure 5.7.

Layer 2: The Digital PCS Data Link Layer 83

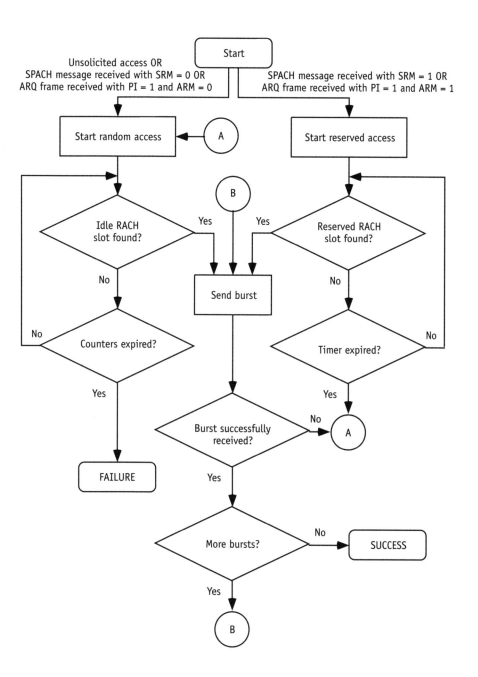

Figure 5.7 Flowchart of mobile access on the RACH.

For unsolicited access attempts and for access attempts in response to a received SPACH message indicating that the next access attempt should be contention-based, the mobile begins the random access process. The mobile monitors the forward DCCH SCF field for an indication that the corresponding RACH slot is idle. When an idle RACH slot is indicated, the mobile attempts the access on that slot. The mobile then monitors the SCF field for an indication that the access attempt was successful. If the access attempt was successful and the mobile has more bursts (layer 2 frames) to send, the mobile transmits in the appropriate RACH slot. If the access attempt was not successful, the mobile reattempts the access up to a maximum number of times allowed by the base station. This process continues until all bursts have been sent, at which time the random access procedure is successfully terminated. If the mobile cannot successfully send or the base station cannot successfully receive all the bursts that the mobile is attempting to send before internal counters in the mobile expire, the random access process is terminated with a failure.

For access attempts in response to a received SPACH message indicating that the next access attempt should be reservation-based, the mobile begins the reserved access process. The mobile monitors the SCF field for an indication that the corresponding RACH slot is reserved for it. If a reserved RACH slot indication is found before an internal time expires, the mobile attempts the access on that slot. Otherwise, the mobile attempts a random access as described previously. After a reserved access attempt, the mobile monitors the SCF field for an indication that the access attempt was successful. If the access attempt was successful and the mobile has more bursts (layer 2 frames) to send, the mobile transmits in the appropriate RACH slot. If the access attempt was not successful, the mobile reattempts the access up to a maximum number of times allowed by the base station. This process continues until all bursts have been sent, at which time the reserved access procedure is successfully terminated. If the mobile cannot successfully send or the base station cannot successfully receive all the bursts that the mobile is attempting to send before internal counters in the mobile expire, the reserved access process is terminated with a failure.

As described above, shared channel feedback plays an important role in the random and reserved access processes. The Partial Echo field indicates to the mobile if the Received/Not Received (R/N) field applies to its last access attempt. The Busy/Reserved/Idle (BRI) field indicates whether the next RACH slot corresponding to this forward DCCH slot is busy, reserved, or idle. This correspondence between RACH and forward DCCH slots is made possible by the definition of subchannels on the RACH. There are six RACH subchannels on a full-rate DCCH, as depicted in Figure 5.8. The shared channel feedback in a F-DCCH slot corresponds to a RACH slot 640 ms behind and ahead of this slot.

5.2.4 ARQ mode operation

ARQ mode is used for acknowledged delivery of layer 3 information to a mobile. A mobile enters ARQ mode when it receives an ARQ mode BEGIN frame on the SPACH with an MSID matching its own. If the Polling Indicator field in the ARQ mode BEGIN frame indicates that the base station desires the mobile to acknowledge receipt of the frame, the mobile sends an ARQ STATUS frame on the RACH. The ARQ STATUS frame is sent using either reserved or random access, based on the value of the ARQ Response Mode field.

Each time the mobile receives an ARQ mode CONTINUE frame, the received frame number is marked as received within the frame number map (FRNO MAP) field. The FRNO MAP field is 26 bits long, so it can support ARQ mode transactions up to 27 ARQ frames long (1 BEGIN and 26 CONTINUE). Each time an ARQ mode CONTINUE frame is received with the Polling Indicator set, the mobile must respond with an ARQ STATUS frame that includes the latest FRNO MAP field. This provides feedback to the base station as to the frames that the mobile has and has not received. The base station may then selectively retransmit the missing frames.

Although the mobile is required to respond to every ARQ mode CONTINUE frame for which the Polling Indicator is set, the network should judiciously select the interval for polling the mobile. Setting the Polling Indicator on every frame would reduce the traffic capacity of the RACH and cause the mobile to buffer potentially many ARQ STATUS frames for transmission.

86 Understanding Digital PCS

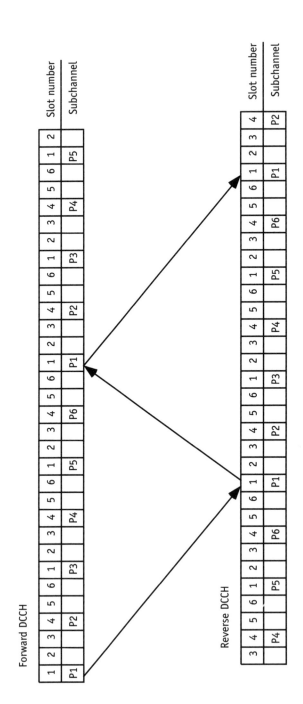

Figure 5.8 Subchannels on the RACH.

5.2.5 Monitoring of radio link quality on the DCCH

The mobile estimates the radio link quality of the DCCH at layer 2. This is accomplished by measuring the word errors based on CRC check failures on the mobile's assigned PCH subchannel. The mobile updates an internally stored monitoring of ratio link quality (MRLQ) counter each time there is a CRC check on the assigned PCH subchannel. The MRLQ counter is initialized to 10 when the mobile obtains service on the DCCH. Each time the CRC check is unsuccessful, the MRLQ counter is decremented by 1, and each time the CRC check is successful it is incremented by 1. The MRLQ counter value is truncated at 10.

When the MRLQ counter reaches 0, the mobile declares a radio link failure on the DCCH. At this time, the mobile proceeds to reselect to a new control channel, as described in Chapter 9.

5.3 Layer 2 of the DTC

Layer 2 of the digital traffic channel is not described to the same level of detail as layer 2 of the DCCH, for the reasons outlined earlier in this chapter. For the DTC, layer 2 comprises the method of supervising the connection on the DTC. This link supervision is enabled through the use of the digital verification color code (DVCC). The DVCC is used to distinguish between the currently assigned DTC and cochannel DTCs. The mobile and the base station perform DVCC verification to ensure they are communicating properly with each other.

The DVCC that the mobile should expect to see on the assigned DTC is provided upon channel assignment. If the mobile reads a DVCC value other than the expected one, this is an indication of cochannel interference, and a five-second fade timer is started in the mobile. If the fade timer expires before the mobile reads the expected DVCC, the call is dropped and the mobile returns to the control channel.

The mobile is required to transpond the DVCC provided to it at channel assignment. The network can then perform a similar DVCC check to ensure radio link continuity. Most networks perform a DVCC verification at initial channel assignment and as part of the handoff process to guarantee that the appropriate mobile is being served on the desired channel.

The DVCC is eight bits in length, yielding 255 distinct values (0 is not used). Service providers typically assign the same DVCC to every DTC in a sector of a cell. In this way, the DVCC also serves as a cell identifier. As described in Chapter 4, channel coding is added to the DVCC to create the coded DVCC (CDVCC), which is transmitted on the DTC.

Further supervision of the DTC link is provided at layer 3 on the fast associated control channel (FACCH) and the slow associated control channel (SACCH). The FACCH and SACCH protocols are described in Chapter 4. This supervision includes physical layer control, channel quality measurements, and handoff orders.

References

[1] Jain, B., and A. Agrawala, *Open Systems Interconnection: Its Architecture and Protocols*, New York: McGraw-Hill, 1993, pp. 83–96.

[2] Telecommunications Industry Association, *TIA/EIA Interim Standard: TDMA Cellular/PCS—Radio Interface—Mobile Station-Base Station Compatibility, Revision A, TIA/EIA/IS-136-A*, Oct. 1996.

Layer 3: The Digital PCS Network Layer

6.1 Introduction to layer 3

Layer 3, or the network layer, provides a means to establish, maintain, and terminate connections between the mobile and the network. It includes a description of the mobile states and procedures required to perform these functions, and defines the message set used for connections between the mobile and the network. The layer 3 mobile states and procedures are described in Chapter 8, after an introduction to the remainder of the network in Chapter 7. Understanding the network architecture is critical to understanding the mobile states and procedures. This chapter is a summary of the layer 3 message set used in Digital PCS.

The layer 3 message set includes messages on both the DCCH and the DTC that are sent across the air interface inside layer 2 protocol frames, as described in Chapter 5 [1]. All layer 3 messages are composed of *information elements*. An information element is a portion of layer 3 data that contains a particular type of information. Once an information element is defined, it may be used in different layer 3 messages. Information elements are the building blocks of messages.

Every layer 3 message begins with the Protocol Discriminator (PD) and Message Type (MT) information elements. The PD information element identifies the layer 3 protocol used for the remainder of the message. Currently, there is only one value defined for the PD—TIA/EIA-136. In the future, another layer 3 protocol could be standardized to fit within a layer 3 message. The second common information element, the MT, identifies the function of the layer 3 message and how the remainder of the information elements are to be interpreted. Figure 6.1 illustrates the common format of layer 3 messages and how they fit into layer 2 frames.

Information elements are identified within a message as either mandatory or optional. When the receiving end of a layer 3 message reads the message type information element, it knows to expect a certain set of mandatory information elements to follow. Every optional information element begins with a Parameter Type field that uniquely identifies the optional information element of the particular message. On the DTC, a remaining length information element always precedes the first optional information element to inform the receiving end of the number of octets that follow. This allows the receiving end to skip the remainder the message if it finds an optional information element it cannot recognize. A layer 3 message is padded with trailing zeros as necessary to align with an octet boundary.

On the DCCH, optional information elements are defined as either comprehension required or comprehension not required, based on the value in the Parameter Type field. If the receiving end of a layer 3 message sees a comprehension required optional information element that it does not recognize, it ignores the remainder of the message. If the optional information element is of the comprehension-not-required variety, it may ignore this information element if it does not recognize it and continue reading the rest of the message. All comprehension-not-required optional information elements must contain a Length Indicator field. This ensures that if the receiving end cannot comprehend the information

Layer 3: The Digital PCS Network Layer 91

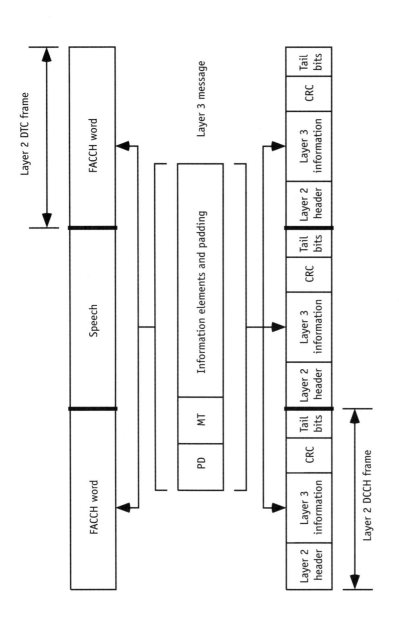

Figure 6.1 Mapping of layer 3 messages into layer 2 frames.

element, it knows how many bits to skip in order to begin looking for an information element it can comprehend.

The rules for handling optional information elements allow for the expansion of layer 3 messages as new features and functionality are standardized. They ensure backward compatibility to mobiles and networks that cannot interpret the new information.

6.2 DCCH message set

6.2.1 BCCH messages

The key BCCH messages (both F-BCCH and E-BCCH) broadcast on a DCCH are described in this section. The messages containing the most time-critical information for a mobile station are placed on the F-BCCH. This ensures that a mobile has the opportunity to read the information every 640 ms. Information on the E-BCCH may extend over many hyperframes and may take the mobile much longer to read. BCCH messages are also identified as either mandatory or optional. If information in a BCCH message must be provided to a mobile for the mobile to operate correctly, then it is mandatory that the message be broadcast.

- *DCCH Structure*: This mandatory F-BCCH message indicates to the mobile the number of slots per superframe dedicated to the different logical channels on the forward DCCH, and identifies if this is a half-rate or full-rate DCCH. Other information provided to the mobile in the DCCH structure message includes the digital verification color code (DVCC), the maximum supported paging frame class, and the number of slots the mobile must examine for a page.

- *Access Parameters*: This mandatory F-BCCH message provides the mobile with information required to access the RACH, including the access power (MS_ACC_PWR), the access burst size (normal or abbreviated), the maximum number of times an individual burst within an access attempt may be transmitted, the maximum number of access attempts allowed, the maximum size of a mobile-originated R-DATA message allowed, and whether or not the mobile should use equalization on the DCCH. The Access Parameters message also indicates to the mobile if it must send an

Authentication message and a Serial Number message coincident with certain other RACH messages.

- *Control Channel Selection Parameters*: This mandatory F-BCCH message provides the mobile with information necessary to determine if the DCCH is suitable for obtaining service, and to perform reselection from the DCCH to a more suitable control channel. The parameters used in reselection that are included in the control channel selection parameters message include SS_SUFF, RSS_ACC_MIN, SCANINTERVAL, and DELAY (see Chapter 9 for a description of the use of these parameters in reselection).

- *Registration Parameters*: This mandatory F-BCCH message identifies to the mobile the type of registrations required on the DCCH. These registration types include power-up (PUREG), power-down (PDREG), new system (SYREG), location area (LAREG), forced (FOREG), and periodic. The registration parameters message also indicates if deregistration (DEREG) is allowed, and if the mobile is allowed to attempt to register with private or residential system IDs (PSIDs/RSIDs) broadcast on the DCCH but not stored in the mobile.

- *System Identity*: This mandatory F-BCCH message identifies the system ID (SID), network type (public, private, and/or residential), the protocol version, the PSIDs/RSIDs available on the DCCH, the mobile country code, and the alphanumeric system ID. The mobile uses this information to determine if the DCCH is suitable from a service perspective, and to identify the most appropriate system to register on when obtaining service. The alphanumeric system ID is a text indication of up to 15 characters in length that can be displayed on the mobile to provide the user with an indication of the service provider from which the mobile has obtained service.

- *Service Menu*: The Service Menu is an optional message that may be broadcast on either the F-BCCH or the E-BCCH. Due to its importance in identifying the types of services available from the DCCH, it is typically placed on the F-BCCH. Among other things, the Service Menu identifies to the mobile the types of vocoders that may be used in the sector or cell. The vocoder information is important for the mobile to know when originating a call from the DCCH. Because the mobile is allowed to originate a call after reading a full cycle of the F-BCCH, the mobile may not know what vocoder to use for the origination unless the Service Menu is placed on the F-BCCH.

- *SOC/BSMC Identification*: The system operator code (SOC)/base station manufacturers code (BSMC) identification message may be broadcast on either the F-BCCH or the E-BCCH. Like the Service Menu message, the SOC/BSMC Identification message is important to the mobile and is typically placed on the F-BCCH. The SOC provided in this message is used by the mobile while performing the intelligent roaming procedures (see Chapter 15) to identify the service provider.

- *Neighbor Cell*: This mandatory E-BCCH message identifies the neighboring control channels for mobile reselection purposes. The reselection parameters for each neighbor are also included in this message. Up to 24 neighboring DCCHs or ACCs can be identified in the Neighbor Cell message. The message is placed on the E-BCCH due to the potential for a sizable message length with the maximum number of neighbors.

- *Regulatory Configuration*: This mandatory E-BCCH message indicates the RF channel allocation for the system. The regulatory configuration identifier (RCI) information present in this message provides Digital PCS with the flexibility to fit into unconventional spectrum allocations and to inform the mobile of this allocation. The mobile may use this information while searching for other DCCHs in the service area.

- *Mobile Assisted Channel Allocation*: This message is an option on either the F-BCCH or the E-BCCH. It is typically placed on the E-BCCH because mobile assisted channel allocation (MACA) is not as high priority as other mobile functions. The MACA message provides a list of up to 15 channels for the mobile to perform signal strength measurements on and report the results back to the base station. The message also indicates if the mobile should report signal strength measurements and bit error rate and word error rate information on the current DCCH. The MACA message further identifies the specific events that should trigger the mobile to send a MACA report.

6.2.2 SPACH messages

The key SPACH messages transmitted on a DCCH are described in this section. Most of the SPACH messages may be sent only on one of the SPACH subchannels—either the SMSCH, the PCH, or the ARCH.

- *Analog Voice Channel Designation*: This message sent on the ARCH is used to assign the mobile to an AVC. Critical information contained in the message includes the AVC channel number, the supervisory audio tone (SAT) to be transponded on the AVC, the transmit power level required of the mobile on the AVC (called the voice mobile attenuation code, or VMAC), and the protocol version of the system.

- *Audit Order*: This message sent on the PCH is used to solicit an audit confirmation message from the mobile. The network may audit the mobile to obtain a MACA report or to force the mobile to reregister.

- *Base Station Challenge Order Confirmation*: This message is sent on the ARCH in response to a Base Station Challenge Order message from the mobile and contains authentication algorithm input. The information contained in the message is AUTHBS, which verifies that the base station is authentic and can update the mobile's shared secret data (SSD) (see Chapter 16).

- *Digital Traffic Channel Designation*: This message sent on the ARCH is used to assign the mobile to a DTC. Critical information contained in the message includes the DTC channel number and time slot, the DVCC to be transponded on the DTC, the transmit power level required of the mobile on the DTC (called the digital mobile attenuation code, or DMAC), the protocol version of the system, time alignment information for mobile transmissions, the vocoder to be used on the DTC, and whether the mobile should use an equalizer on the DTC.

- *Directed Retry*: This message sent on the ARCH is used to force a mobile to reject the current DCCH and reattempt to access an alternate control channel from its neighbor list. It is typically transmitted when the network does not have available resources to set up the call in the current sector or cell.

- *Message Waiting*: This message sent on the PCH is used to inform the mobile that a message is waiting. The mobile may be informed that a voice, short message service (SMS), or fax message is waiting. The number of messages waiting may also be provided to the mobile in this message.

- *R-DATA*: This message sent on the SMSCH is used to convey point-to-point teleservice layer messages to the mobile. Examples of point-to-point teleservices that are conveyed using the R-DATA message include SMS and over-the-air activation teleservice (OATS). The R-DATA message includes a transaction identifier to uniquely identify each message for acknowledgement purposes. It may also include addressing information for the teleservice server network element (see Chapter 7) and the originating user.

- *R-DATA ACCEPT*: This message sent on the SMSCH is used to acknowledge and accept an R-DATA message sent by the mobile. It includes the transaction identifier (R-Transaction Identifier) for the R-DATA message that is being acknowledged. It may also include a delay parameter (R-DATA Delay) that instructs the mobile to wait a specified time before transmitting another R-DATA message. This delay can be used to reduce congestion on the RACH, if required.

- *R-DATA REJECT*: This message sent on the SMSCH is used to acknowledge and reject an R-DATA message sent by the mobile. In addition to the R-Transaction Identifier and the optional R-DATA Delay parameters described for the R-DATA accept message, the R-DATA reject message includes a cause value for the rejection (R-Cause).

- *Registration Accept*: This message is sent on the ARCH in response to a Registration message and is used to notify the mobile that its registration has been completed. The message may optionally include the assignment of a paging frame class to the mobile, information on the current location area, the assignment of a temporary mobile station ID or a user group, and a list of PSIDs and/or RSIDs that the mobile may used in the service area.

- *Registration Reject*: This message is sent on the ARCH in response to a Registration message and is used to notify the mobile that its registration attempt has failed. The message is typically sent when the network cannot verify the authenticity of the mobile. The message includes a cause for the reject, and may include an indication of the length of time the mobile must wait before attempting another registration.

- *Reorder/Intercept*: This message sent on the ARCH is used to inform the mobile that an Origination or R-DATA message sent by the mobile has been rejected, typically due to congestion in the network. The Reorder/Intercept message includes the cause for the reject and the tone indicator for alerting the mobile user to the reject.

- *SPACH Notification*: This message sent on the PCH is used to inform the mobile that the network intends to deliver a message to it on the ARCH or SMSCH. It includes a SPACH Notification Type information element identifying the type of message the mobile should expect to receive next from the network.

- *SSD Update Order*: This message is sent on the PCH or ARCH and causes the mobile to execute the authentication algorithm to update the SSD in the mobile (see Chapter 16). It includes the RANDSSD parameter that is input into the algorithm used to calculate the new value of SSD.

- *Test Registration Response*: This message is sent on the ARCH to inform the mobile whether it is likely to receive service upon attempted registration with a PSID or RSID. It is also used to provide alpha tags associated with PSIDs and RSIDs (see Chapter 14).

- *Unique Challenge Order*: This message sent on the PCH causes the mobile to execute the authentication algorithm and send an

Authentication message to the network. It includes the RANDU parameter that is input into the algorithm to calculate an authentication response (see Chapter 16).

6.2.3 RACH messages

The key RACH messages transmitted on a DCCH are described in this section. Some of these messages may be concatenated and sent by the mobile in a single access attempt, thereby allowing more efficient use of the RACH. RACH messages that may be concatenated are shown in Table 6.1.

- *Audit Confirmation*: This message is sent by the mobile in response to an Audit Order message from the network.

- *Authentication*: This message is used to provide the network with information required to authenticate the mobile (see Chapter 16). The Authentication message is typically transmitted coincident with a Registration, Page Response, Origination, or SPACH Confirmation message.

Table 6.1
Rules for RACH Message Concatenation

Primary Messages	Secondary Messages			
	Authentication	Serial Number	MACA Report	Capability Report
Audit Confirmation	No	No	Yes	No
Base Station Challenge Order	No	Yes	No	No
Origination	Yes	Yes	Yes	No
Page Response	Yes	Yes	Yes	No
Registration	Yes	Yes	Yes	Yes
R-DATA	Yes	Yes	No	No
SPACH Confirmation	Yes	Yes	Yes	No

- *Base Station Challenge Order*: This message is used to confirm the authenticity of the network when it attempts to change the mobile's SSD (see Chapter 16).

- *Capability Report*: This message is used to provide the network with information concerning the capabilities supported by the mobile. This includes the protocol version, station class mark, maximum supported paging frame class, asynchronous data and group 3 fax support, supported frequency bands, vocoder support, and teleservice support.

- *MACA Report*: This message is used to provide signal strength measurements on the channels included in the Mobile Assisted Channel Allocation message on the BCCH, and signal strength, bit error rate, and word error rate on the current DCCH. The Mobile Assisted Channel Allocation message indicates the conditions under which the MACA Report message is transmitted to the network. It may be transmitted coincident with a Registration, Page Response, Origination, or Audit Confirmation message.

- *Origination*: This message is sent when the mobile desires to initiate a voice or data call. The message includes the called party number, the type of service desired (for example, analog speech, digital speech, data, or fax), and the vocoder if digital speech is used.

- *Page Response*: This message is sent in response to a page by the network. It includes essentially the same information as provided in the Origination message, except for the called party number.

- *R-DATA*: This message is used to convey point-to-point teleservice layer messages to the network. The R-DATA message includes a transaction identifier to uniquely identify each message for acknowledgement purposes. It may also include addressing information for the teleservice server network element (see Chapter 7) and the user destination.

- *R-DATA ACCEPT*: This message is used to acknowledge and accept an R-DATA message sent by the network. It includes the transaction identifier (R-Transaction Identifier) for the R-DATA message that is being acknowledged.

- *R-DATA REJECT*: This message is used to acknowledge and reject an R-DATA message sent by the network. It includes a cause value for the rejection (R-Cause).

- *Registration*: This message is used to request registration on the network.

- *Serial Number*: This message is used to provide the network with the electronic serial number (ESN) of the mobile for identification and authentication purposes.

- *SPACH Confirmation*: This message is used to confirm receipt of a SPACH Notification message.

- *SSD Update Order Confirmation*: This message is sent to the network to confirm a successful or failed attempt to update the SSD in the mobile (see Chapter 16).

- *Test Registration*: This message is sent by the mobile to the base station to inquire whether it is likely to receive service should it attempt to register on any given PSID or RSID.

- *Unique Challenge Order Confirmation*: The message is sent in response to a Unique Challenge Order message and includes authentication information to verify the authenticity of the mobile.

6.3 DTC message set

6.3.1 SACCH and FACCH messages on the forward DTC

The key SACCH and FACCH messages sent on the forward DTC are summarized in this section. Some of the messages may only be sent on one of the channel types. High-priority messages are typically sent on the FACCH.

- *Alert with Info*: This message is sent on the FACCH to cause audible or visual signaling to the mobile user related to the initiation of a

Layer 3: The Digital PCS Network Layer   101

call. The calling party number and calling party name can be provided to the mobile in this message.

- *Audit*: This message is sent on the SACCH to determine if a mobile is active in the system. The mobile must acknowledge receipt of the audit message with a Mobile Ack message.

- *Base Station Ack*: This message acknowledges certain messages sent to the network from the mobile. The currently defined messages that are acknowledged are the Connect, Release, Status, and Service Request messages. All of these messages are acknowledged on the FACCH. For the case of a Base Station Ack for a Release message, the Digital Control Channel Information information element may be sent to the mobile to direct the mobile to a DCCH.

- *Base Station Challenge Order Confirmation*: This message is sent on the FACCH in response to a Base Station Challenge Order message from the mobile and contains authentication algorithm input. The information contained in the message is AUTHBS, which verifies that the base station is authentic and can update the mobile's SSD (see Chapter 16).

- *Capability Update Request*: This message is sent on the FACCH to solicit the protocol and service capabilities of the mobile.

- *Capability Update Response*: This message is sent on the FACCH in response to a Capability Update Request message from the mobile and indicates the protocol and service capabilities of the network.

- *Dedicated DTC Handoff*: This message is sent on the FACCH to order the mobile from one DTC to another DTC. It is similar to the handoff message, but may be used for *interhyperband* handoff from 800 MHz to 1,900 MHz, and vice versa.

- *Flash with Info*: This message is sent on the FACCH to convey message waiting information or the calling party number to the mobile.

- *Flash with Info Ack*: This message is sent on the FACCH to acknowledge receipt of a Flash with Info message from the mobile.

- *Handoff*: This message sent on the FACCH orders the mobile from one DTC to another DTC or an AVC. The message includes all the information the mobile needs to perform the handoff, including the channel and time-slot indicator, the SAT or DVCC, the power level to transmit at on the new channel (DMAC or VMAC), time alignment information, and a shortened burst indicator to indicate if the mobile should use shortened burst on the new DTC.

- *Maintenance*: This message is sent on the FACCH in order to check the operation of the mobile. The mobile must acknowledge receipt of the maintenance message with a Mobile Ack message.

- *Measurement Order*: This message is sent on the FACCH to inform the mobile to begin channel quality measurements and reporting on the current channel and a list of up to 24 channels provided in the message.

- *Physical Layer Control*: This message sent on the FACCH is used to control the physical layer of the mobile, including the output power level and time alignment.

- *R-DATA, R-DATA ACCEPT, and R-DATA REJECT*: These messages may be sent on the FACCH or SACCH and are described in Section 6.2.2.

- *Reauthentication Order*: This message is sent on the FACCH and causes the mobile to execute authentication procedures as part of over-the-air activation (see Chapter 12).

- *Release*: This message sent on the FACCH is used to inform the mobile that the currently established call is terminated. It may include the Digital Control Channel information element to direct the mobile to a DCCH.

- *Service Response*: This message is sent on the FACCH in response to a Service Request message from the mobile and indicates that an in-call service change has been accepted or rejected.

- *SSD Update Order*: This message sent on the FACCH and causes the mobile to execute the authentication algorithm to update the SSD in the mobile (see Chapter 16). It includes the RANDSSD parameter that is input into the algorithm used to calculate the new value of SSD.

- *Stop Measurement Order*: This message sent on either the SACCH or the FACCH directs the mobile to stop channel quality measurements and reporting.

- *Unique Challenge Order*: This message sent on the FACCH causes the mobile to execute the authentication algorithm and send a Unique Challenge Order Confirmation message to the network. It includes the RANDU parameter that is input into the algorithm to calculate an authentication response (see Chapter 16).

6.3.2 SACCH and FACCH messages on the reverse DTC

The key SACCH and FACCH messages sent on the reverse DTC are summarized in this section. As is the case on the forward DTC, some of the reverse DTC messages may only be sent on one of the channel types. High-priority messages are typically sent on the FACCH.

- *Base Station Challenge Order*: This message is sent on the FACCH and used to confirm the authenticity of the network when it attempts to change the mobile's SSD (see Chapter 16).

- *Capability Update Request*: This message is sent on the FACCH in order to solicit the protocol and service capabilities of the network.

- *Capability Update Response*: This message is sent on the FACCH in response to a Capability Update Request message and indicates the protocol and service capabilities of the mobile.

- *Connect*: This message sent on the FACCH is used to indicate a call answer by the mobile user.

- *Channel Quality Message*: There are four channel-quality messages sent on the FACCH or SACCH, and each contains the measurement reports for up to six of the RF channels solicited by the network in the measurement order message.

- *Flash with Info*: This message is sent on the FACCH to indicate the user desires to invoke a special service. A Feature Indicator information element identifies the desired service. Features that are invoked with this message include call hold and three-way calling.

- *Flash with Info Ack*: This message is sent on the FACCH to acknowledge a Flash with Info message from the network.

- *Measurement Order Ack*: This message sent on the FACCH acknowledges the start of the channel-quality measurement in the mobile.

- *Mobile Ack*: This message acknowledges certain messages sent to the mobile from the network. Key messages that are acknowledged with a Mobile Ack message are the Alert with Info, Stop Measurement Order, Release, Maintenance, Audit, Handoff, Dedicated DTC Handoff, and Service Response. The Stop Measurement Order and Audit messages may be acknowledged on either the FACCH or the SACCH, while the remainder must be acknowledged on the FACCH.

- *Physical Layer Control Ack*: This message sent on either the FACCH or the SACCH is used to acknowledge the Physical Layer Control message.

- *R-DATA, R-DATA ACCEPT, and R-DATA REJECT*: These messages may be sent on the FACCH or SACCH and are described in Section 6.2.3.

- *Reauthentication Order Confirmation*: This message is sent on the FACCH and contains the response to the Reauthentication Order message. This message is used as part of over-the-air activation (see Chapter 12).

- *Release*: This message sent on the FACCH informs the network that a call currently established is terminated.

- *SSD Update Order Confirmation*: This message on the FACCH is sent to the network to confirm a successful or failed attempt to update the SSD in the mobile (see Chapter 16).

- *Service Request*: This message is sent on the FACCH to request an in-call service change. An example of an in-call service change is if the mobile user desires to turn on voice privacy during a call.

Layer 3: The Digital PCS Network Layer **105**

- *Unique Challenge Order Confirmation*: The message is sent on the FACCH in response to a Unique Challenge Order message and includes authentication information to verify the authenticity of the mobile.

Reference

[1] Telecommunications Industry Association, *TIA/EIA Interim Standard: TDMA Cellular/PCS—Radio Interface—Mobile Station-Base Station Compatibility, Revision A, TIA/EIA/IS-136-A*, Oct. 1996.

Network Architecture and Intersystem Operation

7.1 Network overview

In the Digital PCS standard, the cellular system is composed of the mobile and the network. A mobile is called a mobile station (MS), and the network is composed of the base station (BS), mobile switching center (MSC), and interworking function (IWF) [1]. The BS, MSC, and IWF are collectively called the BMI. The MS and BMI communicate over the TIA/EIA-136 air interface. Internal communications of the BMI are not standardized. BMIs communicate with each other using the TIA/EIA-41 network interface. This simple network reference model is shown in Figure 7.1.

The BMI is actually composed of more network elements than the acronym would suggest. Another network reference model, shown in Figure 7.2, provides a more detailed description of the BMI [2]. In

108   Understanding Digital PCS

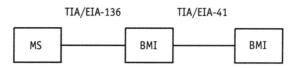

Figure 7.1   Simple Digital PCS network reference model.

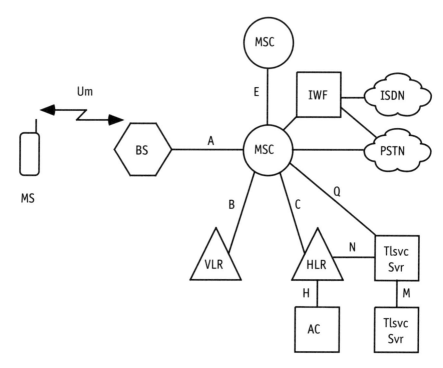

Figure 7.2   Detailed Digital PCS network reference model.

addition to the MS, BS, MSC, and IWF, this model includes the following network elements: home location register (HLR), visitor location register (VLR), authentication center (AC), and teleservice server. Interfaces between the network elements and between the MSC and the public switched telephone network (PSTN) are identified in the model.

The network reference model is a *functional model* because each network element serves a different function. A *physical model* of the network might look much different from Figure 7.2 because different network elements may be combined into the same hardware. For example, the MSC and VLR are often combined, as are the HLR and AC. The network reference model is general enough to apply to many different hardware implementations, however, which makes it a useful tool for explaining the operation of a Digital PCS network. The network elements and interfaces are described in the remainder of this chapter.

7.2 Mobile station

An MS is composed of a control unit, a transceiver unit, and an antenna unit. These units are functional entities and may be housed in the same physical enclosure (for example, a handset). The control unit provides the interface between the user and the cellular network. It may include a microphone, a speaker, a display, a keypad, and other physical controls. The transceiver unit contains logic circuitry and RF transmission and reception circuitry. The logic circuitry encodes and decodes voice and data signals, and the RF transmission and reception circuitry provides modulation, demodulation, and amplification of voice and data signals. The antenna unit connects to the transceiver unit and consists of a vertically polarized broad-bandwidth antenna.

The transceiver unit implements most of the layer 1, 2, and 3 functionality of the Digital PCS standard. Only a minimum of the control unit and antenna unit functionality is standardized. This allows MSs to be differentiated based on usability, performance, and purpose. For example, an MS may be designed small enough to fit into a pocket, or it may be composed of multiple larger units for installation into a vehicle. It may be a traditional mobile phone or a data-only device such as a pager. It may be a stand-alone unit or designed into another consumer product, such as a personal computer. Almost any MS with a transceiver unit conforming to the Digital PCS standard may be produced.

A Digital PCS MS is typically more complex than an AMPS MS. Most Digital PCS MSs are dual-mode (TDMA and AMPS) [3], and many are dual-band (800 MHz and 1,900 MHz). The layer 3 operation of a Digital PCS MS allows for more features and functions than an AMPS MS. For example, a Digital PCS MS may implement an *intelligent roaming* algorithm to automatically select the best service provider from which to obtain service in any location (see Chapter 15). In AMPS, the MS typically defaults to a particular band regardless of the service provider. Some Digital PCS MSs may contain a *short message entity* to send and receive alphanumeric messages, and others may contain sophisticated software supporting *thin client architecture* to enable users access to the World Wide Web, as described in Chapter 12. These and other aspects of Digital PCS MSs are covered in greater detail through the remainder of this text.

7.3 Base station

A BS is composed of a radio transceiver controller, one or more radio transceivers, and an antenna system. The radio transceiver controller multiplexes signals from the radio transceivers to the MSC, and demultiplexes signals from the MSC to the radio transceivers. Each radio transceiver contains logic circuitry and RF transmission and reception circuitry, and performs functions similar to the MS transceiver unit. Some of the layer 2 and 3 functions may be distributed between the radio transceivers, the radio transceiver controller, and other network elements. For example, the MSC may perform the voice coding and decoding operation and an IWF may implement most of layer 2 for data services. The antenna system typically consists of combiners, amplifiers, filters, and transmit and receive antennas to amplify, radiate, and receive signals across the air interface.

The radio transceiver controller manages the radio resources and speech trunks within a sector or cell and performs call-processing functions in conjunction with the MSC. The radio transceiver controller may also monitor and gather statistical information about the operation of the BS and perform diagnostic tests on the BS equipment. It is typically composed of one or more processors, sometimes fully redundant to increase fault tolerance, and interfaces to the MSC and the radio transceivers. Voice and data links connect the radio transceiver controller to the MSC.

The layer 1 functionality of the Digital PCS air interface is implemented in the radio transceivers. The radio transceivers perform timeslot framing and modulate and demodulate voice and data signals. In some implementations of BSs, the radio transceivers have the flexibility to transmit and receive both AMPS and TDMA signals. Other implementations require different radio transceivers for the different technologies. Regardless, a radio transceiver can take on different functions. For example, a TDMA radio transceiver can transmit and receive a full-rate DCCH on two time slots while transmitting and receiving two full-rate DTCs on the remaining time slots.

The antenna system combines the transmissions from multiple radio transceivers, amplifies and filters them, and radiates them through (typically) one transmit antenna. It also receives signals from MSs through one or more receive antennas, amplifies and filters them, and provides the received signals to the radio transceivers for demodulation. The transmit and receive bandpass filters are typically implemented in a duplexer, which affords the radio transceivers with a level of isolation between the transmitted and received signals. Directional transmit and receive antennas may be used for sectorization and increased gain, or omnidirectional antennas may be used for nonsectorized cells. Two receive antennas are typically used to provide diversity reception of signals. Combining the received signals from the two antennas can provide significant diversity gain in the receive path. Additionally, so-called *smart antennas* capable of highly directional beamforming can be used to reduce interference on the uplink from the mobile and increase gain on the downlink to the mobile [4].

7.4 Mobile switching center

The MSC provides the same functionality as a central office switch in the PSTN, and is additionally responsible for call processing, mobility management, and radio resource management [5]. Switching includes providing for the physical means of voice and data transmissions over wireline trunks. Call processing encompasses the functions of call establishment, maintenance, and release to and from mobile users. Mobility management includes all of the processes that enable a user to be mobile. Radio resource management includes assigning radio channels and managing handoffs. The distinction between these functions sometimes blurs, but the MSC is the central controller of them all. The MSC may also perform operations, administration, and maintenance (OA&M) functions.

The call-processing functions of an MSC for a particular mobile begin when the mobile signals the origination of a call or an incoming call is received for a mobile via a voice trunk from the PSTN, and end when the call is completed. The mobility management functions of an MSC for a particular mobile begin when the mobile signals a registration, and end when the mobile powers down or deregisters from the system. The MSC interacts with other MSCs, the HLR, the VLR, BSs, and the PSTN as part of the call-processing and mobility management functions. It uses network signaling and wireline trunks to communicate with these network elements.

Radio resource management includes the functions to establish, maintain, and release radio connections. It includes selecting the best radio channel and call mode to serve a mobile at call establishment and throughout the call. The entire handoff process is part of radio resource management. Maximization of transmission quality through power control, handoff, and other means are key goals of radio resource management.

OA&M functions allow the service provider to control the operation of the cellular network, extract billing and traffic data, and detect and recover from faults. Through the OA&M function, the service provider assigns RF channels and trunks to BSs and sets network parameters. The OA&M function provides calling party number, called party number, chargeable call duration, and other information that is used by the service provider's billing system. It gives the service provider the ability to gather usage statistics, call-blocking data, call completion percentage, blocked call percentage, channel quality statistics, and other measures that may be used to enhance the design of the network. Finally, the OA&M function alerts the service provider to faults in the network and allows the service provider to test different parts of the network.

7.5 Interworking function

An IWF is an entity that performs protocol conversion operations. In the context of Digital PCS, the IWF typically refers to the function that performs the protocol conversion between the IS-130/135 asynchronous data and fax protocols and wireline modem protocols, which is described in Chapter 13. This is necessary because the Digital PCS vocoders are not optimized to pass wireline modem tones. The IS-130/135 protocols are therefore used to transport asynchronous data and fax service from the mobile to the IWF, bypassing the vocoder. The IWF establishes the wireline call and converts the IS-130/135 protocols to a wireline modem protocol such as V.34. The IWF may also provide direct access to a packet data network and the integrated services digital network (ISDN). An IWF may be used to perform conversion for other Digital PCS protocols, such as 136+ packet data (see Chapter 20).

The IWF may be integrated with the MSC or separate from it. When an intersystem handoff occurs during the delivery of asynchronous data or fax service, connectivity to the serving IWF is typically maintained to avoid modem retraining and to maintain continuous data connectivity.

An IWF may also perform protocol conversion, database mapping, and transaction management between different types of cellular networks. For example, an IWF may be used to link the TIA/EIA-41 network to the GSM or PCS1900 network for roaming purposes. This function is particularly useful for service providers who desire to support roaming between the different technologies using multimode mobiles.

7.6 Home location register and visitor location register

The HLR is the primary database that stores information about mobile users. It contains a record for each home subscriber that includes the mobile directory number, subscriber features, subscriber status, and current location information. When a mobile attempts to register on a system, the serving MSC queries the HLR for the mobile to obtain feature information and to notify the HLR of the mobile's whereabouts. When a call is received at the home MSC for a mobile user, the MSC queries the HLR for that mobile to obtain the current status and location of the mobile. TIA/EIA-41 network signaling is used for this communication between the MSC and the HLR. An HLR may be integrated with an MSC, or may stand alone. A stand-alone HLR has the advantage of being able to serve many MSCs, even from different vendors.

The VLR is a database function local to an MSC that maintains temporary records associated with nonhome mobiles registered for service. Upon mobile registration, the VLR temporarily stores subscriber information obtained from the HLR through network signaling. The serving MSC accesses the VLR to retrieve information for the handling of calls to and from the mobile user. When the mobile registers in another location or a timer expires, the VLR record for the mobile user is purged.

The HLR and VLR are primarily databases that store mobile user information. The HLR must also have logic to communicate with MSCs via TIA/EIA-41. There must be a user interface into the HLR for the service provider to provision, or activate, a mobile user for service. There is often also a user interface into the VLR to view the current status of visiting mobiles. This may be used for fraud prevention, gathering information about roamers, and troubleshooting reported problems. The functions of the HLR and VLR are central to the mobile operation described in Chapter 8; the voice, teleservice, and data services described in Chapters 11, 12, and 13, respectively; and the authentication process described in Chapter 16.

7.7 Authentication center

The AC is responsible for managing authentication information related to a mobile, including the authentication key (A-Key) and shared secret data (SSD) for a mobile. It also executes the cellular authentication and voice encryption (CAVE) algorithm to generate SSD, perform authentication, and calculate the voice privacy mask and signaling message encryption key for a particular mobile. The AC may be integrated with the MSC, HLR, or both. It may also be stand-alone. All network communications with the AC is through the HLR. The AC may also have a data link to an authentication management system, or may have the functionality built-in, to obtain valid A-Key/ESN pairs from mobile manufacturers. The functions of the AC are described in greater detail in Chapter 16.

7.8 Teleservice server

The teleservice server may also be called the message center. However, the teleservice server performs the processing of different types of teleservices other than short messages, so the more general name is appropriate. The teleservice server formats, stores, and forwards teleservice layer messages to and from a mobile. It uses network signaling to communicate with other network entities to locate a mobile and deliver a teleservice layer message. Example of teleservices that a teleservice server may support include cellular messaging teleservice (CMT), over-the-air activation teleservice (OATS), and over-the-air programming teleservice (OPTS). A window into the teleservice server to develop new teleservices may be provided through a service creation environment.

Different functional entities may reside in a teleservice server to support different teleservices. For example, CMT is provided by the short message service center (SMSC) function, and OATS and OPTS are provided by the over-the-air activation function (OTAF). These functions may be implemented in software in the same platform and may use the same teleservice message storage database and network interface. The teleservice server may communicate with other network entities through Signaling System No. 7 (SS7), X.25, or the Internet Protocol (IP).

The teleservice server takes messages delivered from a short message entity, formats them into a TIA/EIA-41 message, and transports the message to the serving MSC if the mobile is active. If the mobile is not active, the teleservice server stores the message for the mobile until it becomes available for service again. The service provider may specify the maximum length of time that messages may be stored in the teleservice server. The TIA/EIA-41 protocol allows the user data portion of a single teleservice message segment to be approximately 200 octets in length. The teleservice server may apply teleservice segmentation and reassembly (TSAR) to send messages longer than 200 octets in length. The teleservice server may also apply compression and encryption to the message, provided the destination is aware of the method to decompress and decrypt the message.

A teleservice server may serve one or many cellular markets. Teleservice servers may communicate with each other to deliver teleservice messages between mobiles. The teleservice server may also be linked to an interactive voice response (IVR) system or a paging network to provide additional teleservice delivery mechanisms. The teleservice server may therefore need to provide interworking between many different protocols. Chapter 12 describes the functions of the teleservice server and elaborates on the teleservices identified above.

7.9 Network interfaces

Many of the interfaces between the network elements are defined by the TIA/EIA-41 standard. TIA/EIA-41 is the intersystem signaling protocol that allows mobile users to roam between different networks. TIA/EIA-41 provides for standard communication between the following network elements: MSC-VLR, MSC-HLR, HLR-VLR, MSC-MSC, HLR-AC, MC-HLR, and MC-MSC. The interface between the BS and the MSC, called the A interface, is not necessarily standardized for Digital PCS. As a result, these two network elements are typically provided by the same infrastructure supplier.

The physical, data link, and network layers of TIA/EIA-41 are typically provided by the SS7 protocol, although X.25 may be used instead. The TIA/EIA-41 SS7-based protocol structure is illustrated in Figure 7.3. SS7 is an out-of-band common channel signaling protocol that uses packets to convey signaling information between two entities through multiple switching nodes [6]. SS7 provides what is called the data transfer services in TIA/EIA-41. The TIA/EIA-41 SS7-based data transfer services comprise the SS7 message transfer part (MTP) and signaling connection control part (SCCP). MTP provides the physical, data link, and portions of the network layer protocols. SCCP provides the remainder of the network layer protocol. Only SCCP class 0 connectionless service is used in TIA/EIA-41. This means that signaling between entities is provided through datagrams containing source and destination address information.

The data transfer services support the TIA/EIA-41 application services, which consist of the remaining layers in the OSI reference

Network Architecture and Intersystem Operation 117

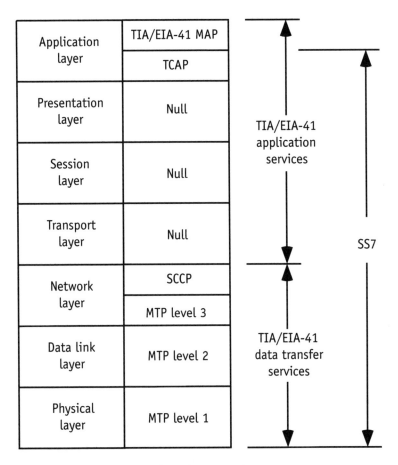

Figure 7.3   TIA/EIA-41 SS7-based protocol structure.

model—the transport, session, presentation, and application layers. The application services distinguish TIA/EIA-41 from other uses of SS7 and make possible the transfer of application information between network elements in the TIA/EIA-41 network reference model. TIA/EIA-41 application services comprise the SS7 transaction capabilities application part (TCAP) and the TIA/EIA-41 mobile application part (MAP). TCAP defines standard message formats (called components) and message sequences (called packages) used by the TIA/EIA-41 MAP. The TIA/EIA-41 MAP defines the actual messages and operations associated with the messages in the various network elements.

Through the protocol described above, TIA/EIA-41 realizes the functions of automatic roaming, authentication, call processing, intersystem handoff, and teleservice delivery. Automatic roaming allows mobile users to obtain service outside the home service provider area. Authentication prevents fraudulent mobiles from accessing a network. Call processing includes the functions to establish, maintain, and release calls to and from mobile users. Intersystem handoff allows a mobile to move from one radio channel to another while maintaining a call. Teleservice delivery allows a mobile to send and receive data such as short messages. Together, these functions enable the network to provide service to mobile users.

References

[1] Telecommunications Industry Association, *TIA/EIA Interim Standard: TDMA Cellular/PCS—Radio Interface—Mobile Station-Base Station Compatibility, Revision A, TIA/EIA/IS-136-A*, Oct. 1996.

[2] Gallagher, M. D., and R. A. Snyder, *Mobile Telecommunications Networking with IS-41*, New York: McGraw-Hill, 1997, pp. 33–57.

[3] Harte, L., *Dual Mode Cellular*, Bridgeville, PA: P. T. Steiner Publishing Co., 1992, pp. 5–1 to 5–46.

[4] Winters, Jack H., "Smart Antennas for Wireless Systems," *IEEE Personal Communications*, Vol. 5, No. 1, Feb. 1998, pp. 23–27.

[5] Boucher, N. J., *The Cellular Radio Handbook*, Third Edition, Mill Valley, CA: Quantum Publishing, 1995, pp. 277–313.

[6] Lin, Y., and S. K. DeVries, "PCS Network Signaling Using SS7," *IEEE Personal Communications*, Vol. 2, No. 3, June 1995, pp. 44–55.

8
Digital PCS Mobile Operation

8.1 Mobile types

There are different types of Digital PCS mobiles to meet the needs of different users and applications. The types of mobiles may be differentiated by their power class and suite of features and functionality. The suite of features and functionality of a mobile defines its tier. There are different tiers of mobiles to appeal to different users.

The allowable power classes of a Digital PCS mobile were described in Chapter 4. The majority of Digital PCS mobiles are class IV handhelds. This means that they are capable of transmitting at a maximum nominal power (effective radiated power) of 28 dBm (approximately 600 mW) and stepping down in 4-dB steps to a nominal power of -4 dBm (approximately 0.4 mW) at the low end. When installed in a vehicle, a booster and external antenna may be used to change the mobile to a class I mobile with a maximum nominal output power of 36 dBm (approximately 4W).

A class I mobile is useful in rural areas with long distances between cell sites and where signal strength may be very low in some areas.

Digital PCS mobiles are often split into tiers to appeal to different users, in much the same way as automobiles. A low-tier Digital PCS mobile may provide basic voice service, a limited set of advanced features, average standby and talk times, and few accessory choices. A mid-tier Digital PCS mobile may be smaller in size than the low-tier mobile, but perhaps with a larger display. It may provide more advanced services and better standby and talk times through a more efficient standard battery. A high-tier Digital PCS mobile may be either a voice-centric mobile that is very small in size or a data-centric mobile with a large screen and much functionality, such as a personal communicator. Regardless of the tier, all Digital PCS mobiles operate essentially the same because they must conform to the TIA/EIA-136 standard for mobile states and procedures.

8.2 Mobile states

A mobile's operation is most easily explained through a description of its states and processes. These are defined in the TIA/EIA-136 standard, so that every Digital PCS mobile will operate essentially the same way and have similar performance characteristics [1]. Since the operation of a mobile in AMPS mode is well known [2], this description will focus on the operation of a Digital PCS mobile in digital mode.

Figure 8.1 shows a state diagram of a Digital PCS mobile. This model is implemented in software in a Digital PCS mobile. There are various procedures that a mobile must execute while in a state. These procedures are described in the next section. A mobile will leave a particular state and enter another state based on rules contained in the TIA/EIA-136 standard. The states are described in the remainder of this section.

8.2.1 Control Channel Scanning and Locking

A mobile is in the Control Channel Scanning and Locking state when it is searching for service. The mobile enters this state when powered on, at the end of a call, when transitioning from an ACC to a DCCH, or when a

Digital PCS Mobile Operation   121

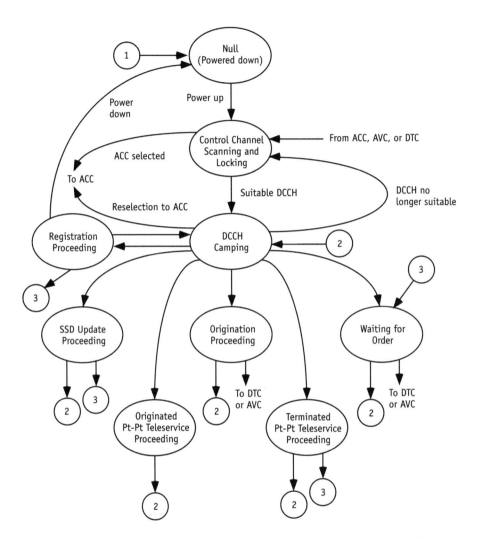

Figure 8.1 State diagram of a Digital PCS mobile.

DCCH is deemed to be unsuitable for camping and there are no suitable neighbors for reselection. A mobile executes the DCCH Scanning and Locking procedure while in this state, attempting to find a DCCH. At any time while in the Control Channel Scanning and Locking state, a mobile may determine that an ACC is preferred over a DCCH and execute the

Initialization task searching for an ACC. If a mobile finds a DCCH suitable for camping while executing the DCCH Scanning and Locking procedure, it enters the DCCH Camping state.

A mobile is not available to place or receive calls while in the Control Channel Scanning and Locking state. A mobile typically provides a no service indication to the user while searching for service to indicate that it is not available for use. Minimizing the time that a mobile spends in the Control Channel Scanning and Locking state is a goal of the mobile designer and the service provider. In addition to being out of service when in this state, a mobile also must expend more battery power than desirable to scan for service.

8.2.2 DCCH Camping

The DCCH Camping state is the normal state of a mobile while in service on a DCCH and not processing a transaction. The mobile leaves the DCCH Camping state to process any of the following transactions: registration, call origination, call termination, teleservice termination or origination, and shared secret data (SSD) updating. A condition that precipitates leaving the DCCH Camping state and returning to the Control Channel Scanning and Locking state is losing service on the current DCCH and the inability to find another control channel through reselection. The mobile also leaves the DCCH Camping state if the mobile reselects to an ACC.

A mobile must read a full cycle of the fast broadcast control channel (F-BCCH) and extended broadcast control channel (E-BCCH) upon entering the DCCH Camping state on a new DCCH. This is necessary for the mobile to obtain information critical for operation on the DCCH, including access and registration parameters. A mobile may make an access attempt on the DCCH after reading the F-BCCH and before reading the E-BCCH, however. This allows a mobile to originate a call more quickly, but does not allow the mobile to identify neighboring control channels for reselection purposes.

The mobile periodically executes a number of processes and procedures while in the DCCH Camping state, including paging channel (PCH) subchannel monitoring and control channel reselection. A mobile remains in the DCCH Camping state while it reselects from one DCCH to another. A mobile may enter sleep mode (see Chapter 10) while in the DCCH Camping state on a DCCH and not executing any other procedures.

8.2.3 Registration Proceeding

A mobile moves to the Registration Proceeding state after sending a Registration or Test Registration message on a DCCH. It is in this state while waiting for a response from the network to its registration attempt. It leaves this state under the following conditions: (1) upon determining through layer 2 feedback mechanisms that the Registration message was received by the network, and this is a power-down registration; (2) upon determining through layer 2 feedback mechanisms that the Registration message was not received by the network; (3) upon receiving a Registration Accept, Registration Reject, or Test Registration message from the network; (4) upon receiving a page or SPACH Notification message from the network; or (5) upon expiration of a timer indicating the time to wait between successfully sending a Registration message and receiving a response from the network.

When a mobile enters the Registration Proceeding state after sending a Registration or Test Registration message, it waits to receive a layer 2 access success indication. If a success indication is received, the mobile starts the DEREG_TMR if a deregistration condition exists, or the REG_TMR if a registration or test registration condition exists. Under normal circumstances, the network will respond with a Registration Accept, Registration Reject, or Test Registration Response message before the timer that is running expires. The network will send a Registration Accept in response to a registration if it has verified the authenticity of the mobile. If the authenticity of the mobile has not been verified, the network will send a Registration Reject. The network will send a Test Registration Response message in response to a Test Registration message from the mobile. The Test Registration Response indicates the private and residential system identities (PSIDs and RSIDs) the mobile is allowed service on, and provides the alpha tags for the PSIDs and RSIDs supported by the current DCCH.

If the mobile has sent a power-down registration that has been received by the network, it proceeds to complete the power-down process and enter the Null state. If the mobile receives a page or SPACH Notification addressed to it while in the Registration Proceeding state, it moves to the Waiting for Order state. For all other cases in which it leaves the Registration Proceeding state, the mobile returns to the DCCH Camping state.

8.2.4 Origination Proceeding

A mobile is in the Origination Proceeding state after sending an Origination message on a DCCH and while waiting for a response from the network. It leaves this state under the following conditions: (1) upon determining through layer 2 feedback mechanisms that the Origination message was not received by the network; (2) upon receiving a Digital Traffic Channel Designation, Analog Voice Channel Designation, Directed Retry, or Reorder/Intercept message from the network; (3) upon expiration of a timer indicating the time to wait between successfully sending an Origination message and receiving a response from the network; or (4) upon detecting a power-down condition.

If the mobile determines that the Origination message was not received by the network, it returns to the DCCH Camping state, where it may reinitiate the origination process. If the mobile receives a Digital Traffic Channel Designation or Analog Voice Channel Designation message, it moves to the assigned channel. If the mobile receives a Directed Retry message, it invokes the Control Channel Reselection procedure in an attempt to find a better control channel from which to originate. If one is found, the mobile originates on this new control channel. If one is not found, it returns to the DCCH Camping state. If the mobile receives a Reorder/Intercept message from the network because the network does not recognize its identity, the mobile may resend the Origination message using another of its identities (if it has not already attempted an origination with this alternate identity). If the mobile does not have another identity for which an origination has not already been attempted, or for any other reorder/intercept code, the mobile returns to the DCCH Camping state. The mobile also returns to the DCCH Camping state if the origination timer expires and the mobile has not received a response from the network. If a power-down condition is detected, the mobile invokes the Registration procedure, sends a power-down registration if the network indicates support for power-down registration, and enters the Registration Proceeding state.

8.2.5 Waiting for Order

A mobile is in the Waiting for Order state after sending a Page Response message on a DCCH and while waiting for a response from the network. It leaves this state under the following conditions: (1) upon determining through layer 2 feedback mechanisms that the Page Response message was not received by the network; (2) upon receiving a Digital Traffic Channel Designation, Analog Voice Channel Designation, Directed Retry, or Release message from the network; (3) upon expiration of a timer indicating the time to wait between successfully sending a Page Response message and receiving a response from the network; or (4) upon detecting a power-down condition.

If the mobile determines that the Page Response message was not received by the network, it returns to the DCCH Camping state. If the mobile receives a Digital Traffic Channel Designation or Analog Voice Channel Designation message, it moves to the assigned channel. If the mobile receives a Directed Retry message, it invokes the Control Channel Reselection procedure in an attempt to find a better ontrol channel. If one is found, the mobile resends the Page Response on this new control channel. If one is not found, it returns to the DCCH Camping state. If the mobile receives a Release message from the network, the mobile returns to the DCCH Camping state. The mobile also returns to the DCCH Camping state if the Waiting for Order timer expires and the mobile has not received a response from the network. If a power-down condition is detected, the mobile invokes the Registration procedure, sends a power-down registration if the network indicates support for power-down registration, and enters the Registration Proceeding state.

8.2.6 Terminated Point-to-Point Teleservice Proceeding

A mobile is in the Terminated Point-to-Point Teleservice Proceeding state after sending a SPACH Confirmation message in response to a SPACH Notification message from the network. It leaves this state under the following conditions: (1) upon determining through layer 2 feedback mechanisms that the SPACH Confirmation message was not received by the network; (2) upon receiving an R-DATA message from the network; (3) upon receiving a page, SPACH Notification, Digital Traffic Channel Designation, or Release message from the network; (4) upon expiration of a timer indicating the time to wait between successfully sending a SPACH Confirmation message and receiving a response from the network; or (5) upon detecting a power-down condition.

If the mobile determines that the SPACH Confirmation message was not received by the network, it returns to the DCCH Camping state. If the mobile receives an R-DATA message addressed to it, it acknowledges receipt of the message by sending either an R-DATA ACCEPT or R-DATA REJECT message to the network. If the mobile receives a page message addressed to it, it invokes the Termination procedure, sends a Page Response message, and enters the Waiting for Order state. If the mobile receives a Digital Traffic Channel Designation message, it moves to the assigned channel. If the mobile receives a Release message from the network, the mobile returns to the DCCH Camping state. The mobile also returns to the DCCH Camping state if the timer expires and the mobile has not received a response from the network. If a power-down condition is detected, the mobile invokes the Registration procedure, sends a power-down registration if the network indicates support for power-down registration, and enters the Registration Proceeding state.

8.2.7 SSD Update Proceeding

A mobile is in the SSD Update Proceeding state if it has (1) sent a SPACH Confirmation message in response to a SPACH Notification message indicating SSD update, (2) sent a Base Station Challenge Order message in response to an SSD Update Order message, or (3) sent an SSD Update Order Confirmation message. It leaves this state under the following conditions: (1) upon determining through layer 2 feedback mechanisms that the message it sent was not received by the network, (2) upon determining through layer 2 feedback mechanisms that the SSD Update Order Confirmation message was successfully received by the network, (3) upon receiving a page message from the network, (4) upon expiration of a timer indicating the time to wait between successfully sending either a SPACH Confirmation or Base Station Challenge Order message and receiving a response from the network, or (5) upon detecting a power-down condition.

When a mobile in the DCCH Camping state receives a SPACH Notification indicating SSD update, it sends a SPACH Confirmation and enters the SSD Update Proceeding state. If it receives a layer 2 access success indication, it sets a timer called the SPACH_TMR and remains in the SSD Update Proceeding state. If the mobile receives an SSD Update Order message before the SPACH_TMR expires, it begins the SSD update process by executing the CAVE algorithm (described in Chapter 16) and sends the network a Base Station Challenge Order message. If it receives a layer 2 access success indication after sending the Base Station Challenge Order, it sets a timer called the SSDU_TMR and remains in the SSD Update Proceeding state. If the mobile receives a Base Station Challenge Order Confirmation message, the mobile processes the confirmation as described in Chapter 16. If the network has proved its authenticity, the mobile updates its SSD. Otherwise, the mobile keeps its old SSD. The mobile then sends an SSD Update Order Confirmation message to the network indicating if its SSD has been updated and remains in the SSD Update Proceeding state until a layer 2 access indication has been received.

If a layer 2 access failure indication is received, a layer 2 access success indication is received for an SSD Update Order Confirmation, or either SPACH_TMR or SSDU_TMR expires before receiving a response, the mobile returns to the DCCH Camping state. If the mobile receives a page message addressed to it anytime it is in the SSD Update Proceeding state, it invokes the Termination procedure, sends a Page Response message, and enters the Waiting for Order state. If a power-down condition is detected while in the SSD Update Proceeding state, the mobile invokes the Registration procedure, sends a power-down registration if the network indicates support for power-down registration, and enters the Registration Proceeding state.

8.2.8 Originated Point-to-Point Teleservice Proceeding

A mobile is in the Originated Point-to-Point Teleservice Proceeding state after it sends an R-DATA message to the network but has not yet received a response from the network. It leaves this state under the following conditions: (1) upon determining through layer 2 feedback mechanisms that the R-DATA message was not received by the network, (2) upon receiving an R-DATA ACCEPT or R-DATA REJECT message from the network, (3) upon receiving a Reorder/Intercept message from the network, (4) upon expiration of a timer indicating the time to wait between successfully sending an R-DATA message and receiving a response from the network, or (5) upon detecting a power-down condition.

If the timer expires before a response to the R-DATA message is received from the network, the mobile may resend the R-DATA message a second time. If a power-down condition is detected, the mobile invokes the Registration procedure, sends a power-down registration if the network indicates support for power-down registration, and enters the Registration Proceeding state. For all other conditions, the mobile returns to the DCCH Camping state.

8.3 Mobile station procedures

8.3.1 DCCH Scanning and Locking

The mobile executes the DCCH Scanning and Locking procedure in an attempt to find a DCCH while in the Control Channel Scanning and Locking state. The mobile may have precise information on the frequency and partial DVCC of a DCCH that it has obtained from the Control Channel Information message on the ACC or from the DCCH information provided on call release from a DTC or AVC. It uses this information while executing the DCCH Scanning and Locking procedure to more rapidly find a DCCH. If this type of information is not available to the mobile, it uses the Intelligent Roaming procedure to search for a DCCH, as described in Chapter 15. If a mobile finds a DCCH while executing the DCCH Scanning and Locking procedure, it invokes the Control Channel Selection procedure to determine if the DCCH is suitable for service.

While executing the Intelligent Roaming procedure within the DCCH Scanning and Locking procedure, the mobile uses historic information on the location of DCCHs to speed up the scanning process. If the mobile is not successful in finding a suitable DCCH from historic information, it searches for DCCHs in a prescribed order within any given frequency band, based on the likelihood of finding a DCCH. In the 800-MHz band, channels are scanned in probability blocks, and in the 1,900-MHz band, channels are scanned in sub-bands. The probability block and sub-band allocations are described in Chapter 17.

It is worth noting that prior to TIA/EIA-136-A, a Digital PCS mobile was not required to search for DCCHs according to the Intelligent Roaming procedure described in Chapter 15. Instead, the mobile could employ any nonstandard algorithm as long as it resulted in finding a control channel. Most algorithms started with a historic search of DCCHs used in the past. If a DCCH was not found from this historic search, the mobile would scan the ACCs in an attempt to find control channel information pointing to a DCCH (assuming the scanning was conducted in the 800-MHz band where there are ACCs). If no ACCs were found, the mobile would conduct a band scan in order of probability blocks searching for a DCCH. If an ACC was found, but no control channel information was present, the mobile would obtain service on the ACC.

8.3.2 Control Channel Selection

A mobile executes the Control Channel Selection procedure from the Control Channel Scanning and Locking state once a DCCH is found using the DCCH Scanning and Locking procedure. The Control Channel Selection procedure is executed in order to determine if the DCCH is suitable for camping. There are three parts to the procedure: Signal Strength Aspects Determination, Service Provider Acceptability Determination, and Service Aspects Determination. The DCCH must pass each of these procedures in order to be deemed suitable for camping. If the mobile determines the DCCH is suitable for camping, it enters the DCCH Camping state. Otherwise, it remains in the Control Channel Scanning and Locking state and again invokes the DCCH Scanning and Locking procedure.

The mobile executes the Signal Strength Aspects Determination procedure to determine if the signal strength on the DCCH is adequate for camping. The mobile measures the received signal strength on the DCCH and reads access threshold information on the F-BCCH. It then compares the received signal strength to the access threshold, as described in

Chapter 17. If the received signal strength is above the access threshold, the DCCH is suitable for camping from a signal strength perspective and the mobile executes the Service Provider Acceptability Determination procedure. Otherwise, the mobile returns to the Intelligent Roaming procedure.

The mobile executes the Service Provider Acceptability Determination procedure to determine if the DCCH belongs to a preferred service provider. The mobile must read system identity and system operator code information from the F-BCCH and compare it to SID and SOC information stored in the mobile to determine if the service provider is acceptable, as described in Chapter 15. If the service provider is acceptable, the mobile executes the Service Aspects Determination procedure. Otherwise, the mobile returns to the Intelligent Roaming procedure.

The mobile executes the Service Aspects Determination procedure to determine if it should be operating on a different DCCH in the current sector, to verify that the DCCH is not barred for service, and to verify that the DCCH supports a network type subscribed to by the mobile. The mobile must read additional messages on the F-BCCH to make these determinations. If the mobile determines that it should be operating on a different DCCH in the same sector, it moves to that DCCH and reenters the Signal Strength Aspects Determination procedure. If the current DCCH is barred or does not support a network type subscribed to by the mobile, the mobile reenters the Intelligent Roaming procedure.

8.3.3 Control Channel Reselection

A mobile executes the Control Channel Reselection procedure to find a better control channel from which to obtain service. It typically invokes this procedure from the DCCH Camping state, although Control Channel Reselection may be invoked from other states if a Directed Retry message is received from the network. As reselection is covered in detail in Chapter 9, this section only summarizes the procedure.

There are two parts to the Control Channel Reselection procedure: Control Channel Locking and Reselection Criteria. In the Control Channel Locking procedure, the mobile measures and averages the signal strength on the current DCCH and all neighboring control channels. It does this at a measurement interval determined by information broadcast on the current DCCH. The signal strength of the current DCCH is measured every SCANINTERVAL. The signal strength of neighboring control channels is measured either every SCANINTERVAL or every other SCANINTERVAL. The mobile keeps a running average of the last five signal strength measurements on the current DCCH and every neighboring control channel. Additionally, it keeps a running average of the last two signal strength measurements on the current DCCH.

The Reselection Criteria procedure is composed of three additional procedures that are invoked sequentially by the mobile: Reselection Trigger Conditions, Candidate Eligibility Filtering, and Candidate Reselection Rules. In the Reselection Trigger Conditions, the mobile determines if a condition exists that would trigger a further evaluation of candidate control channels for reselection. In the Candidate Eligibility Filtering procedure, the mobile executes an algorithm specific to the trigger condition to screen the candidate control channels based on biases and thresholds identified by reading information available on the current DCCH. In the Candidate Reselection Rules procedure, the mobile determines the best candidate control channel, synchronizes to it, reads information on it, and reselects to it if it is deemed a better control channel than the current DCCH. If the best candidate is the current DCCH, the mobile remains on that DCCH. If no suitable control channel can be found after executing the Candidate Reselection Rules procedure, including the current DCCH, the mobile moves to the Control Channel Scanning and Locking state.

8.3.4 Termination

A mobile executes the Termination procedure when it receives any of a number of messages on the DCCH, including but not limited to a page, Audit Order, SPACH Notification, SSD Update Order, Unique Challenge Order, and Message Waiting. The mobile must respond to the message received from the network within a time window specified in the TIA/EIA-136 standard. There are different time windows for different messages.

If the mobile receives a page message with an acceptable service code, it returns a Page Response message and enters the Waiting for Order state. If the mobile receives an Audit Order, it returns an Audit

Confirmation message along with any required coincidental messages (such as a MACA Report) and remains in the DCCH Camping state. If the Audit Order includes a Forced Reregistration flag, the mobile then invokes the Registration procedure. If the mobile receives a SPACH Notification, it responds with a SPACH Confirmation. If the SPACH Notification indicates an SSD update, the mobile enters the SSD Update Proceeding state. If the SPACH Notification indicates an R-DATA message, the mobile enters the Terminated Point-to-Point Proceeding state. If the SPACH Notification indicates MACA, the mobile includes a MACA Report with the SPACH Confirmation and remains in the DCCH Camping state. If the mobile receives an SSD Update Order, it issues a Base Station Challenge Order and enters the SSD Update Proceeding state. If the mobile receives a Unique Challenge Order, it issues a Unique Challenge Order Confirmation message and remains in the DCCH Camping state. If the mobile receives a Message Waiting message, it issues a SPACH Confirmation and remains in the DCCH Camping state.

8.3.5 Origination

A mobile executes the Origination procedure when a user origination is detected. If the origination is an emergency call, the mobile ignores the overload class and sends an Origination message to the network. Otherwise, it must verify that the overload control bit corresponding to its overload class is enabled in the Overload Class message on the F-BCCH (if it is broadcast) before sending the Origination message. The mobile then enters the Origination Proceeding state.

8.3.6 Originated Point-to-Point Teleservice

A mobile executes the Originated Point-to-Point Teleservice procedure when it has a mobile-originated teleservice message to send in an R-DATA message. If the mobile has multiple R-DATA messages to send and the first R-DATA message has been sent, the mobile must delay a length of time specified on the DCCH in the R-DATA Delay parameter before sending the next R-DATA message. The mobile must also verify that the overload control bit corresponding to its overload class is enabled before sending the R-DATA message. The length of the R-DATA message cannot exceed the length specified in the R-DATA Message Length parameter sent on the DCCH. The mobile must also verify that the network supports mobile-originated teleservices before sending an R-DATA message. Once the mobile sends the R-DATA message, it enters the Originated Point-to-Point Teleservice Proceeding state.

8.3.7 Registration-related procedures

There are three primary registration-related procedures: Registration, Registration Success, and Registration Reject. A mobile executes the Registration procedure when it determines a registration may be in order. It may determine this upon power-up; power-down; change in SID, PSID, or RSID; change in control channel type; change in location area; expiration of a periodic registration timer; or receipt of a forced registration indication from the network. Before sending a Registration message, the mobile must verify that the overload control bit corresponding to its overload class is enabled and that the registration parameters associated with the required registration type are set. The mobile must indicate in the Registration message the type of registration being attempted. Upon sending a Registration message, the mobile enters the Registration Proceeding state.

For most networks, REGH and REGR are set to 1 to indicate that both home and roaming mobiles are allowed to register. This will be assumed in the following discussion. If a mobile powers up and finds a DCCH with PUREG set to 1, it sends a Registration message with a registration type of power-up. This indicates to the network that the mobile is available for service. When a mobile detects that the user desires to power-down, the mobile sends a Registration message of type power-down if PDREG is set to 1 on the current DCCH. This indicates to the network that the mobile is unavailable for service. If a mobile decides to change the system (SID, PSID, or RSID) it is obtaining service from on the current DCCH and SYREG is set to 1, the mobile sends a Registration message of type new system. If the mobile transitions from an ACC to a DCCH and SYREG is set to 1 on the DCCH, the mobile sends a Registration message of type new system. Registering upon change in control channel type helps the network to know where to page the mobile.

Upon registration to a DCCH, a mobile stores a list of values called RNUMs that define its current location area. When a mobile reselects to a new DCCH and the RNUM that it broadcasts is not in the mobile's current RNUM list, and LAREG is set to 1 on the new DCCH, the mobile sends a Registration message of type location area. When the network has a message for a mobile, it first pages the mobile in the last location area in which it registered. Many networks are divided into location areas in order to more efficiently page mobiles in this manner.

Many networks require mobiles to periodically register in order to renew their VLR records. If a mobile fails to periodically register, its VLR record may be removed and calls may not be routed to it. The network may broadcast a value called REGPER, which indicates to the mobile the period within which it must register with the network. The mobile keeps a periodic registration timer, and upon expiration of this timer it invokes the registration procedure, sends a Registration message of type periodic registration, and moves to the Registration Proceeding state.

The network may force a mobile to register by sending it an Audit Order with a forced reregistration indication. If this occurs, the mobile must invoke the Registration procedure and send a Registration message of type forced, if FOREG is set to 1 on the DCCH. The network may force a mobile to reregister if it suspects fraudulent use of the mobile.

A mobile executes the Registration Success procedure after receiving a Registration Accept message from the network while in the Registration Proceeding state. The network may provide the mobile with various information in the Registration Accept message, including a list of RNUMs indicating the current location area, a temporary mobile station identity (TMSI) for the mobile to use for addressing purposes, an assigned PFC, and a list of PSIDs and RSIDs for which the mobile is allowed service. If the network sends any of this information in the Registration Accept message, the mobile stores this information and considers itself registered with the system.

A mobile executes the Registration Reject procedure after receiving a Registration Reject message from the network while in the Registration Proceeding state. The mobile takes different actions based on the reject cause received in the Registration Reject message. If the reject cause is unknown MSID and the mobile has other permanent identities that have not previously been rejected, the mobile may attempt to register with one of these alternate identities. If the reject cause is PSID/RSID removal, the mobile removes from memory the PSID or RSID with which it attempted to register. The Registration Reject message may include a reject time, which indicates the length of time within which the mobile may not attempt to re-register. When this time has expired, the mobile may reattempt the registration.

8.3.8 Authentication procedures

A mobile executes the authentication procedures when it is necessary for it to authenticate with the network. It typically invokes these procedures coincident with a registration, origination, or termination on the DCCH. The authentication procedures are also used in updating the mobile's SSD while in the SSD Update Proceeding state. The authentication procedures are described in detail in Chapter 16.

If the AUTH bit is set to 1 on the DCCH upon which the mobile is operating, it must send the Authentication message coincident with a Registration, Origination, Page Response, SPACH Confirmation, or R-DATA message. The Authentication message includes an authentication response computed by the mobile based on mobile-specific information, a secret key, and a random number broadcast on the DCCH. The network verifies the authenticity of the mobile by running the same authentication algorithm and matching the authentication result from the mobile. This verifies that the mobile and the network have the same secret key. If the network computes a different authentication response than the mobile, the network may deny service to the mobile. It may send the mobile a Registration Reject message with reject cause authentication failure, or it may simply block terminations to and originations from the mobile.

The type of authentication described above is known a global challenge authentication because all mobiles must authenticate. The network may also choose to authenticate only certain mobiles it suspects of fraudulent use. It does this by sending the mobile a Unique Challenge Order message. The mobile executes the Unique Challenge-Response procedure upon receipt of this message. It calculates a unique challenge authentication response in a manner similar to that described for the case of a global challenge and sends it to the network in a Unique Challenge Order Confirmation message. The network compares the mobile's response with its own calculation to verify its authenticity.

8.3.9 Mobile Assisted Channel Allocation

If the Mobile Assisted Channel Allocation message is broadcast on the DCCH upon which a mobile is operating, it is required to execute the MACA procedures and make MACA reports to the network. The network may use MACA information reported by mobiles to implement adaptive channel allocation. The MACA procedures are described in detail in Chapter 10.

A mobile may be required to make and report long-term MACA (LTM) and short-term MACA (STM) measurements. LTM measurements include word error rate, bit error rate, and received signal strength on the current DCCH. STM measurements include received signal strength on the current DCCH and a list of channels provided in the Mobile Assisted Channel Allocation message. Depending upon settings in this message, the mobile may be required to send a MACA Report message to the network containing LTM and STM measurements coincident with an Audit Confirmation, Page Response, Origination, and/or a Registration message.

8.4 Mobile station identities

A Digital PCS mobile may have different mobile station identifications (MSIDs) that it uses to communicate with the cellular network. These identities include the mobile identification number (MIN), international mobile station identity (IMSI), and temporary mobile station identity (TMSI). The MIN and IMSI are permanent MSIDs (PMSIDs) that are stored in the mobile's NAM. The TMSI is temporary, as its name implies, and may be assigned and changed at different times during a power cycle of the mobile. A mobile may have a MIN, an IMSI, and a TMSI all at the same time. There are rules that govern the use of each MSID to keep the mobile from becoming schizophrenic. This section describes each type of mobile identification and the rules that apply to their use.

8.4.1 MIN

The MIN is the most well-known means of identifying a mobile. It is the traditional 10-digit directory telephone number that is dialable by another party. The MIN is composed of a three-digit area code (NPA, or

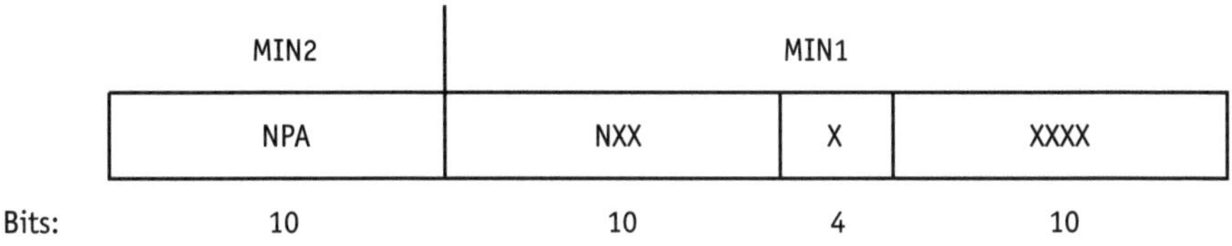
MIN2
MIN1
NPA
NXX
X
XXXX
Bits:
10
10
4
10

Figure 8.2 Coding of 10-digit directory telephone number into 34-bit MIN.

numbering plan assignment), a three-digit exchange office number (NXX), and a four-digit extension number (XXXX). This 10-digit number is coded into a 34-bit binary number, as shown in Figure 8.2. The 24 least significant bits of the 34-bit MIN comprise MIN1, and the 10 most significant bits comprise MIN2.

The MIN may be used to page a mobile or send messages to a mobile on the DCCH or ACC. In fact, the MIN is the only mobile station identification used on the ACC. For this reason, the MIN will remain one of the primary MSIDs for years to come. If a mobile has registered with its MIN and has not been assigned a TMSI, it monitors the control channel for messages addressed to its MIN. On the DCCH, up to three mobiles may be paged at the same time in one SPACH slot with MINs, as described in Chapter 5.

One way to implement wireless number portability is to separate the MIN from the mobile directory number (MDN). The MDN remains the dialable number that does not change when the user changes service providers. The MIN changes as the home service provider changes, so the MIN can still be used to identify the home system. With this method of wireless number portability, the mobile user and the caller do not need to know the mobile's MIN, only the user's MDN. The cellular network performs the translation from the MDN to the MIN for communication with the mobile.

8.4.2 IMSI

The MIN is based on the North American numbering plan and is not well suited for some international applications. The International Telecommunications Union (ITU) has established the IMSI as the PMSID for worldwide use in addressing mobiles. The IMSI uniquely identifies the mobile, the home wireless network, and the home country of the network and the mobile. The IMSI format is based on ITU Recommendation E.212. It is 15 digits in length, composed of a three-digit mobile country code (MCC), a three-digit mobile network code (MNC), and a nine-digit mobile station identification number (MSIN). The IMSI format and encoding for TIA/EIA-136 are shown in Figure 8.3.

Like the MIN, the IMSI may be used to page a mobile or send messages to a mobile on the DCCH. If a mobile has registered with its IMSI and has not been assigned a TMSI, it monitors the control channel for messages addressed to its IMSI. Layer 2 signaling in TIA/EIA-136 only allows one mobile at a time to be paged in a SPACH slot using an IMSI.

Wireless number portability may be implemented with IMSIs. Because the IMSI is not a dialable number in North America, an MDN is typically associated with an IMSI. The MDN is translated into the IMSI for communications with the mobile. When the mobile user changes service providers, the IMSI changes but the MDN does not.

8.4.3 TMSI

TMSIs constitute a powerful method for addressing mobiles. Addressing mobiles with TMSIs leads to a more efficient use of the DCCH and more secure communications over the air interface. When a mobile registers with the network, the network may assign the mobile a TMSI in the Registration Accept message on the DCCH. The mobile then monitors its PCH subchannel for messages addressed to its TMSI instead of to its

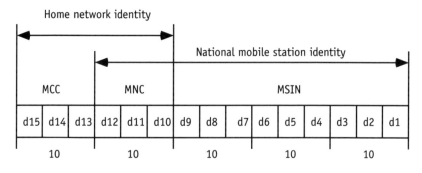

Figure 8.3 IMSI formatting and encoding.

PMSID. The TMSI may be changed by the network every time the mobile reregisters.

Addressing mobiles through TMSIs leads to a more efficient use of the DCCH because more mobiles can be paged with TMSIs than with PMSIDs. This is because the TMSI is shorter than either the MIN or the IMSI. A TMSI can be either 24 or 20 bits in length, compared with a 34-bit MIN and a 50-bit IMSI. This means that four mobiles can be paged in a single SPACH slot using 24-bit TMSIs (called a hard quadruple page), or five mobiles can be paged in a single SPACH slot using 20-bit TMSIs (called a hard penta page). TMSIs are particularly important when the IMSI is used as the primary PMSID for the mobile, because only one mobile using an IMSI can be addressed at a time on a SPACH slot. This significantly reduces the paging capacity on a DCCH unless TMSIs are used.

Addressing mobiles through TMSIs is also more secure than using the PMSID. This is because using TMSIs makes it more difficult for a fraudster to obtain a PMSID/ESN pair over the air, and more difficult to impersonate a mobile when obtaining service. The PMSID is not transmitted except during the registration/registration accept process. A fraudster would have to capture this registration in order to obtain the MIN or IMSI of the mobile. Because the TMSI can be changed by the serving network, and because the TMSI is not used in the authentication process, knowledge of the TMSI by a fraudster is of little value.

References

[1] Telecommunications Industry Association, *TIA/EIA Interim Standard: TDMA Cellular/PCS—Radio Interface—Mobile Station-Base Station Compatibility, Revision A, TIA/EIA/IS-136-A*, Oct. 1996.

[2] Harte, L., *Dual Mode Cellular*, Bridgeville, PA: P. T. Steiner Publishing Co., 1992, pp. 2–1 to 2–34.

9

Reselection and Hierarchical Cell Structures

R ESELECTION IS THE PROCESS that a mobile uses to move seamlessly between control channels [1]. *Hierarchical cell structures* (HCS) may be used by the service provider to guide a mobile to reselect to the most appropriate control channel in any given location. With HCS, the service provider can distribute traffic among cells and sectors as desired to efficiently use available resources. Reselection and HCS together form a powerful tool for the service provider to increase the quality of service offered to the mobile user.

9.1 Neighbor cells and neighbor cell information

Reselection and HCS are only available when a mobile is operating on a DCCH, not an ACC. This is because reselection and HCS require neighbor cell information to be broadcast to mobiles, and this is only provided on the DCCH. Neighbor cell information is broadcast on the E-BCCH of every DCCH in either the Neighbor Cell or Neighbor Cell (Multihyperband) message at layer 3. Upon camping on a DCCH, a mobile reads the E-BCCH and stores neighbor cell information for use in the reselection process. The neighbor cell information used in reselection is summarized in Table 9.1.

Table 9.1
Primary Neighbor Cell Information Used in Reselection

Information Element	Description
SERV_SS	A bias used to increase the preference of the current DCCH compared to the neighboring control channels.
DVCC or DCC	The digital verification color code or digital color code of the neighboring control channel.
RESEL_OFFSET	A bias to increase or decrease the preference of a neighboring control channel.
SS_SUFF	An absolute signal strength threshold that must be met before reselecting to a neighboring control channel of CELLTYPE preferred.
DELAY	A parameter that keeps a mobile from considering a neighboring control channel as a reselection candidate until a time delay has been met.
HL_FREQ	An indicator that sets the measurement interval for the neighboring control channel. If HL_FREQ is set to 1, the signal strength of the neighbor is measured every SCANINTERVAL. If HL_FREQ is set to 0, the measurement interval is 2 × SCANINTERVAL.
CELLTYPE	An identification of the preference of the neighboring control channel, which can be Regular, Preferred, or Nonpreferred.

Reselection and Hierarchical Cell Structures   143

Information Element	Description
Network Type	An identification of the network types (public, private, and/or residential) supported on the neighboring control channel.
Directed Retry Channel	A flag that identifies if the neighboring control channel should be considered a candidate when the mobile receives a directed retry message.
MS_ACC_PWR	The maximum nominal output power that the mobile must use when accessing the neighboring control channel.
RSS_ACC_MIN	The minimum received signal strength required to access the neighboring control channel.

A typical DCCH broadcasts a neighbor cell list containing the control channels of the surrounding cells or sectors, overlay cells or sectors, and underlay cells or sectors. The neighbor cell list may contain both DCCHs and ACCs. However, ACCs are typically only included in the neighbor cell list in border areas. This is because if the mobile reselects to an ACC, the reselection process cannot be used to seamlessly transition to another control channel. Instead, the mobile must either wait for a periodic rescan of the 21 ACCs or lose the current ACC and rescan the 21 ACCs.

Up to 24 neighboring control channels can be broadcast in the Neighbor Cell or Neighbor Cell (Multihyperband) message. Only neighboring control channels in the current hyperband can be included in the Neighbor Cell message, while the Neighbor Cell (Multihyperband) message may include neighbors in both hyperbands. For the case of a 1,900-MHz hyperband system broadcasting the Neighbor Cell message, no ACCs can be included as neighbors. This is because ACCs are not defined for the 1,900-MHz hyperband.

9.2 Control Channel Locking and Reselection Criteria procedures

While in the DCCH Camping state (see Chapter 8), a mobile periodically measures the signal strength on the current DCCH and all control channels in the neighbor cell list, as described in Chapter 10. This measurement process is called the Control Channel Locking procedure, and is executed every SCANINTERVAL. The mobile keeps a running average of the last five signal strength measurements (called Long_RSS) on the current DCCH and the neighboring control channels. The mobile also keeps a running average of the last two signal strength measurements (called Short_RSS) on the current DCCH. A control channel must have a valid Long_RSS prior to being considered for reselection purposes. This requirement sets the minimum time that a mobile must camp on a DCCH before reselecting to a neighboring control channel. Assuming the SCANINTERVAL is set to 1 hyperframe (1.28s) and HL_FREQ is set to 1 for a neighboring control channel, it takes 6.4s (5 × 1.28s) before the mobile may reselect to that neighbor. When the received signal strength on a channel is referred to throughout the remainder of this chapter, it is intended to refer to Long_RSS unless otherwise noted.

The mobile executes the Reselection Criteria procedure if a reselection trigger condition is detected while in the DCCH Camping, Origination Proceeding, or Waiting for Order state. The Reselection Criteria procedure is composed of three distinct and sequential procedures: Reselection Trigger Condition, Candidate Eligibility Filtering, and Candidate Reselection Rules. Each procedure methodically screens and prioritizes the candidate control channels until only one is left—the one from which the mobile should obtain service. It should be noted that this candidate might be the exact same DCCH upon which the mobile is currently operating.

9.3 Reselection Trigger Conditions

There are seven Reselection Trigger Conditions (RTCs), as described in Table 9.2. The Directed Retry condition is the only RTC that the mobile detects outside the DCCH Camping state. The Directed Retry condition is detected within either the Origination Proceeding or Waiting for Order state. When one of the conditions described in Table 9.2 is detected, the mobile invokes the Reselection Criteria procedure.

Table 9.2
Reselection Trigger Conditions

Reselection Trigger Condition	Description
Radio Link Failure	This condition is detected when the word error rate (WER) on the current DCCH becomes excessive. A WER occurs when there is a CRC check failure on a mobile's PCH. A mobile keeps a counter running that decrements every time a WER is declared and increments every time a CRC check is successful on its PCH. The counter is truncated at 10. When the counter reaches 0, a Radio Link Failure condition is declared. This condition is most often caused by interference.
Cell Barred	This condition is detected whenever the mobile either reads a layer 2 go away flag set to 1 or a layer 3 cell barred information element indicating barring. A cell may be marked as barred when the radio supporting the DCCH is taken out for maintenance.
Server Degradation	This condition is detected whenever the signal strength on the current DCCH, as estimated by Long_RSS, has fallen below the minimum required to access the channel, and Short_RSS $\leq$ Long_RSS. This last criterion ensures that the mobile does not reselect if it is coming out of a short-term fade on the current channel. This condition may occur if a mobile turns a corner and the signal strength on the current DCCH falls significantly and rapidly due to shadowing of the signal.
Directed Retry	This condition is detected whenever the mobile receives a Directed Retry message. It may occur when either the mobile user originates a call or a page is received and there are no traffic channels available for assignment in the current sector or cell.
Priority System	This condition is detected whenever the mobile identifies the current control channel as a public service profile (PSP) of an autonomous system, or the nonpublic search procedure results in the identification of a candidate DCCH. For both of these cases, either the mobile or the mobile user decides to acquire service on a DCCH supporting a network type, SID, PSID, or RSID determined to be of higher priority than that supported on the current control channel.

146 Understanding Digital PCS

Table 9.2 (continued)

Reselection Trigger Condition	Description
Service Offering	This condition is detected whenever a mobile decides to evaluate the services offered by other DCCHs. It may occur as the result of the mobile determining that a particular service is not offered on the current DCCH, but is offered on a neighboring one. An example might be asynchronous data and group 3 fax service, which might not be available on an autonomous system but is available on the public cellular network.
Periodic Evaluation	This condition occurs at least once every SCANINTERVAL when a control channel has a valid Long_RSS. This is the most frequently occurring RTC.

For the Directed Retry RTC, only those neighboring control channels marked as directed retry candidates are considered for reselection. When a Priority System condition is declared, the private operating frequencies (POFs) associated with the matching PSP are considered for reselection. For all other RTCs, all the neighboring control channels are considered for reselection.

9.4 Candidate Eligibility Filtering

The Candidate Eligibility Filtering (CEF) procedure is a set of algorithms designed to create a list of candidate control channels that are eligible for reselection. The specific CEF algorithm executed by the mobile depends upon the RTC. Table 9.3 shows the mapping of RTCs into CEFs. Every viable control channel identified in the Reselection Trigger Conditions procedure is evaluated in Candidate Eligibility Filtering.

Figures 9.1 through 9.5 show flowcharts of the five CEF algorithms. The calculations and parameters used by the mobile while executing the CEFs are summarized in Table 9.4. Refer to Table 9.1 for CEF parameters previously defined in this chapter. In the table and the figures that follow, the subscript *cand* refers to a candidate control channel and the subscript *cur* refers to the current DCCH. The function $max(X, Y)$ returns the maximum of X and Y.

Table 9.3
Mapping of RTCs into CEFs

Reselection Trigger Condition	Candidate Eligibility Filter
Radio Link Failure	CEF1
Cell Barred	CEF2
Server Degradation	CEF2
Directed Retry	CEF3
Priority System	CEF2 if SERV_SS = 0
	CEF4 if SERV_SS > 0
Service Offering	CEF4
Periodic Evaluation	CEF5

CEF1 is invoked when a Radio Link Failure condition is declared. All viable candidate control channels are evaluated and those with received signal strength above the minimum required to access the channel are marked as eligible for reselection. There is no delay associated with marking a candidate control channel as eligible for reselection under CEF1. This allows a mobile to immediately reselect to another control channel when a Radio Link Failure occurs on the current DCCH.

CEF2 is invoked when a Cell Barred or Server Degradation condition is declared, or when a Priority System condition is declared and SERV_SS is set to 0 on the current DCCH. A candidate control channel may be marked as eligible for reselection based on either of two criteria. For both criteria, the received signal strength on any candidate must be above the minimum required to access the channel for some amount of time—specified by the DELAY parameter for that neighbor—before declaring the candidate eligible for reselection. Eligibility criteria 1 (called *criteria* in the TIA/EIA-136 standard, not *criterion*) applies if the CELLTYPE of the candidate is Preferred and the received signal strength of the candidate is above an absolute threshold for that candidate (SS_SUFFcand). Candidates that do not pass eligibility criteria 1 are filtered with eligibility criteria 2. For a candidate to be marked as eligible for reselection by criteria 2, the received signal strength on the channel

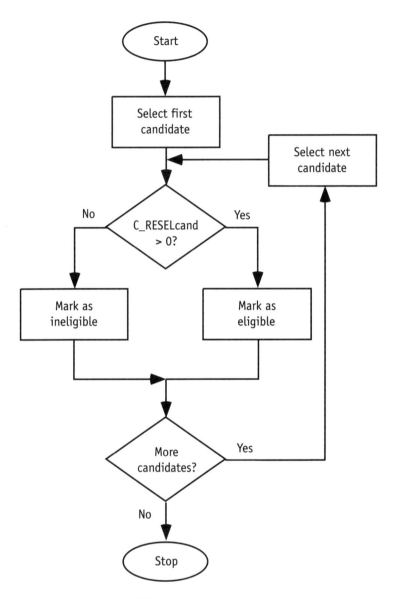

Figure 9.1 Flowchart of CEF1 algorithm.

must be above that of the current DCCH by a relative offset (RESEL_OFFSET, which may be either positive or negative). If the CELLTYPE of the candidate is either Regular or Preferred, this is the

Reselection and Hierarchical Cell Structures 149

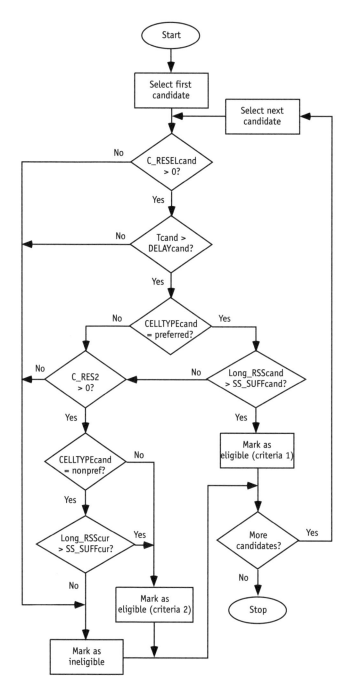

Figure 9.2 Flowchart of CEF2 algorithm.

150 Understanding Digital PCS

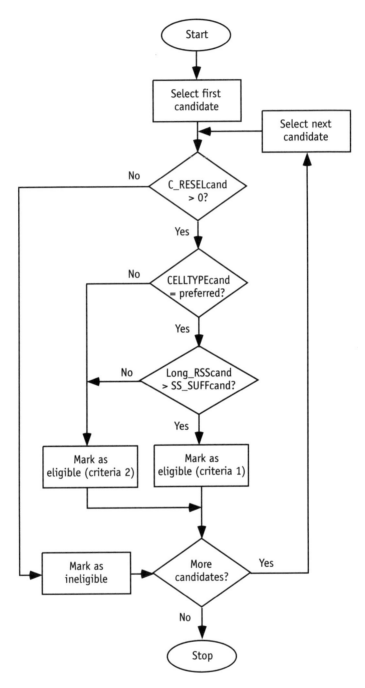

Figure 9.3 Flowchart of CEF3 algorithm.

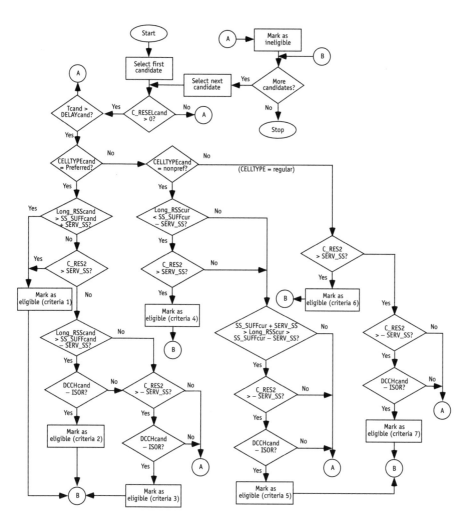

Figure 9.4 Flowchart of CEF4 algorithm.

only additional condition that must be met for the candidate to be declared eligible for reselection by criteria 2. If the CELLTYPE of the candidate is Nonpreferred, the received signal strength on the current DCCH must be below an absolute threshold (SS_SUFFcur) for the candidate to be declared eligible for reselection by criteria 2.

CEF3 is invoked when a Directed Retry condition is declared. It is very similar to CEF1 except that the CELLTYPE of the candidate is

152 Understanding Digital PCS

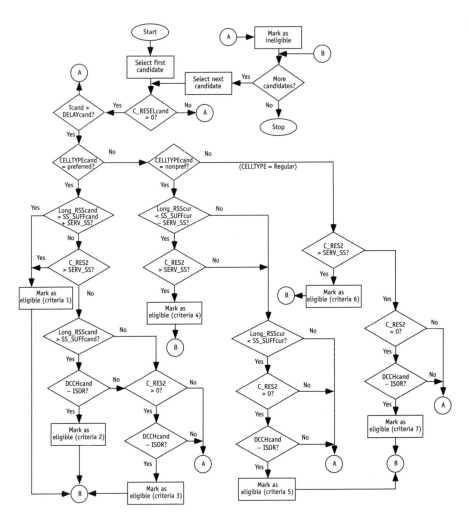

Figure 9.5 Flowchart of CEF5 algorithm.

considered in the filtering process. A candidate control channel may be marked as eligible for reselection based on either of two criteria. For both criteria, the received signal strength on any candidate must be above the minimum required to access the channel. Eligibility criteria 1 applies if the CELLTYPE of the candidate is Preferred and the received signal strength of the candidate is above an absolute threshold for that candidate

Table 9.4
Additional Calculations and Parameters Used in Candidate Eligibility Filtering

Parameter Name	Description

C_RESEL	IF((MS_ACC_PWRcand 4 dBm AND MS_CLASS = 4) OR MS_ACC_PWRcand $\geq$ 8 dBm)
	   C_RESEL = Long_RSS - RSS_ACC_MIN - max(MS_ACC_PWR - P, 0)
	 ELSE
	   C_RESEL = -1

MS_CLASS	The mobile's power class.
P	The maximum nominal output power of the mobile, defined by its power class, in dBm.
C_RES2	C_RESELcand - C_RESELcur + RESEL_OFFSETcand
T	A timer that starts when C_RESELcand becomes greater than 0. It is reset when C_RESELcand becomes less than or equal to 0.
ISOR	Ineligible for service offering reasons, a flag that is set in the Candidate Reselection Rules procedure to indicate that a neighboring control channel does not offer a service desired by the mobile.

(SS_SUFFcand). Candidates that do not pass eligibility criteria 1 are considered to pass by eligibility criteria 2.

CEF4 is invoked when a Service Offering condition is declared, or when a Priority System condition is declared and SERV_SS is greater than 0 on the current DCCH. CEF5 is invoked when a Periodic Evaluation condition is declared. Because the CEF4 and CEF5 algorithms are almost identical, they are described together here. There are seven criteria by which a candidate control channel may be considered eligible for reselection—three apply to Preferred neighbors, two apply to Nonpreferred neighbors, and two apply to Regular neighbors. For all criteria, the received signal strength on any candidate must be above the minimum required to access the channel for some amount of time (DELAYcand) before declaring the candidate eligible for reselection. If the CELLTYPE of the candidate is Preferred, it may be declared eligible for reselection according to one of the following rules:

- If the received signal strength on the candidate is greater than SS_SUFFcand plus SERV_SS, or if the received signal strength on the candidate is above that of the current DCCH by RESEL_OFFSET plus SERV_SS, it is eligible. This is criteria 1.

- Otherwise, if the received signal strength on the candidate is greater than SS_SUFFcand minus SERV_SS, and the candidate has not been marked as ineligible for service offering reasons (ISOR), it is eligible. This is criteria 2.

- Otherwise, if the received signal strength on the candidate is above that of the current DCCH by RESEL_OFFSET minus SERV_SS for CEF4, or RESEL_OFFSET for CEF5, and the candidate has not been marked as ISOR, it is eligible. This is criteria 3.

For CEF4 and CEF5, if the CELLTYPE of the candidate is Nonpreferred, it may be declared eligible for reselection according to one of the following rules:

- If the received signal strength on the current DCCH is less than SS_SUFFcur minus SERV_SS, and if the received signal strength on the candidate is above that of the current DCCH by RESEL_OFFSET plus SERV_SS, it is eligible. This is criteria 4.

- Otherwise, for CEF4, if the received signal strength on the current DCCH is between SS_SUFFcur + SERV_SS and SS_SUFFcur-SERV_SS, the received signal strength on the candidate is above that of the current DCCH by RESEL_OFFSET minus SERV_SS, and the candidate has not been marked as ISOR, it is eligible. This is criteria 5 for CEF4.

- Otherwise, for CEF5, if the received signal strength on the current DCCH is below SS_SUFFcur, the received signal strength on the candidate is above that of the current DCCH by RESEL_OFFSET, and the candidate has not been marked as ISOR, it is eligible. This is criteria 5 for CEF5.

For CEF4 and CEF5, if the CELLTYPE of the candidate is Regular, it may be declared eligible for reselection according to one of the following rules:

- If the received signal strength on the candidate is above that of the current DCCH by RESEL_OFFSET plus SERV_SS, it is eligible. This is criteria 6.

- Otherwise, if the received signal strength on the candidate is above that of the current DCCH by RESEL_OFFSET minus SERV_SS for CEF4, or by RESEL_OFFSET for CEF5, and the candidate has not been marked as ISOR, it is eligible. This is criteria 7.

9.5 Candidate Reselection Rules

The Candidate Reselection Rules procedure is invoked after completion of the Candidate Eligibility Filtering procedure. Those neighboring control channels marked as eligible candidates for reselection as a result of execution of one of the CEF algorithms are evaluated within the Candidate Reselection Rules procedure. There are four subprocedures within the candidate reselection rules procedure: CAND_1 Determination, CAND_1 Examination, Suitable CAND_1 Found, and Suitable CAND_1 Not Found. The end result of a mobile's execution of the Candidate Reselection Rules procedure is (1) reselection to a new control channel, (2) entry into the DCCH Camping state on the current DCCH, or (3) entry into the Control Channel Scanning and Locking state due to the absence of a suitable control channel.

The mobile prioritizes the eligible candidates within the CAND_1 Determination subprocedure and identifies the best candidate control channel based on the CEF and the criteria for which the candidate was marked as eligible for reselection. In general, candidates marked with a CELLTYPE of Preferred take priority over those marked as Regular, and Regular candidates take priority over those marked as Nonpreferred. Within the same CELLTYPE, those candidates having a higher received signal strength generally take priority over those having a lower received signal strength. The highest priority eligible candidate control channel is marked as CAND_1, and the mobile proceeds to the CAND_1

Examination subprocedure. If no CAND_1 is available, the mobile invokes the Suitable CAND_1 Not Found subprocedure.

In the CAND_1 Examination subprocedure, the mobile synchronizes to the CAND_1 control channel, reads broadcast information, and verifies that it is suitable for service. To be suitable, the control channel must be broadcasting the same color code (DVCC or DCC) as the neighbor list entry, and it must not be barred. If this is not the case, the mobile marks the control channel as temporarily ineligible for reselection (TIR). If CAND_1 is marked as TIR and there are more eligible candidate control channels, the mobile returns to the CAND_1 Determination subprocedure and identifies a new CAND_1. If CAND_1 is marked as TIR and there are no more eligible control channels, the mobile invokes the Suitable CAND_1 Not Found subprocedure.

CAND_1 may also be marked as ISOR while the mobile is executing the CAND_1 Examination subprocedure. The control channel is marked as ISOR if it does not offer the service desired by the mobile. This service may be a desired SID, PSID, RSID, or network type. It may also be a service such as asynchronous data and group 3 fax. If the control channel is marked as ISOR, the mobile may (1) terminate the Control Channel Reselection procedure and remain in the DCCH Camping state, (2) restart the Candidate Reselection Rules procedure to identify a new CAND_1 if there are more eligible control channels, or (3) invoke the Suitable CAND_1 Not Found subprocedure if there are no more eligible control channels.

If the mobile verifies that CAND_1 is a suitable control channel while executing the CAND_1 Examination subprocedure, the mobile proceeds as follows:

- If a Directed Retry condition exists, the mobile proceeds to complete the directed retry on CAND_1.

- Otherwise, if CAND_1 is an ACC, the mobile completes the Initialization task on the ACC and enters the Idle task.

- Otherwise, if CAND_1 is a DCCH, the mobile enters the Suitable CAND_1 Found subprocedure.

Reselection and Hierarchical Cell Structures 157

Within the Suitable CAND_1 Found subprocedure, the mobile declares CAND_1 as the current DCCH. For the Suitable CAND_1 Not Found subprocedure, the mobile proceeds as follows:

- If a Directed Retry condition exists, the mobile stops any timers running as a result of the directed retry, terminates the Control Channel Reselection procedure, and enters the DCCH Camping state on the current DCCH.

- Otherwise, if either a Service Offering or Periodic Evaluation condition exists, the mobile terminates the Control Channel Reselection procedure and remains in the DCCH Camping state on the current DCCH.

- Otherwise, the mobile terminates the Control Channel Reselection procedure and enters the Control Channel Scanning and Locking state.

9.6 Reselection examples

The preceding description of the Reselection Criteria procedure gives insight into the algorithmic design of reselection, but may not afford a clear understanding of how reselection actually works. This section provides examples of reselection for all of the reselection trigger conditions except Service Offering. All of the CEFs are exercised in this section, however. For all of the examples, a mobile is in the DCCH Camping state on DCCH1. DCCH1's neighbor list contains both DCCH2, DCCH3, and DCCH4. In all of the examples, it is assumed that the mobile has already performed enough signal strength measurements to build up a valid Long_RSS for all four DCCHs. Key reselection parameters used in the examples to follow are summarized in Table 9.5. It is assumed that DELAY is set to zero in all these examples. The effects of DELAY are considered in the following section. The mobile used in these examples is a class IV mobile capable of transmitting at a maximum power of 28 dBm.

Table 9.5
Reselection Parameters Used in Examples

Parameter	DCCH1	DCCH2	DCCH3	DCCH4
SERV_SS	6	—	—	—
RESEL_OFFSET	—	−2 dB	−4 dB	−6 dB
SS_SUFF	−89 dBm	−89 dBm	−99 dBm	−109 dBm
CELLTYPE	—	Preferred	Regular	Nonpreferred
Network type	Public	Public, Private	Public	Public
Directed retry channel	—	Yes	Yes	No
MS_ACC_PWR	28 dBm	20 dBm	28 dBm	28 dBm
RSS_ACC_MIN	−105 dBm	−101 dBm	−105 dBm	−105 dBm

9.6.1 Radio Link Failure

Excessive word errors are occurring on DCCH1 as a result of cochannel interference. The mobile detects this fact when the word error counter reaches 0, and it declares a Radio Link Failure. The mobile immediately invokes the Candidate Eligibility Filtering procedure. The CEF1 algorithm is executed with DCCH2, DCCH3, and DCCH4 as candidates. The mobile's received signal strengths (Long_RSS) on the DCCHs are DCCH1 = −85 dBm, DCCH2 = −87 dBm, DCCH3 = −87 dBm, and DCCH4 = −87 dBm. Table 9.6 summarizes the calculations the mobile makes during execution of the CEF1 algorithm.

The mobile enters the Candidate Reselection Rules procedure with all three neighboring DCCHs marked as eligible for reselection. DCCH4 is marked as CAND_1 during CAND_1 Determination because it has the highest C_RESEL of any of them. This is because it has the lowest RSS_ACC_MIN. Note that due to the drastic nature of a Radio Link Failure, the CELLTYPE of the candidates, and the DELAY associated with them, are not considered during the Reselection procedure. Note also that the received signal strength on the current DCCH is not a factor in the reselection calculations, even though it is higher than that of any of the candidates. This is because there is something wrong with the

Table 9.6
CEF1 Calculations for a Radio Link Failure Example

Parameter	DCCH2	DCCH3	DCCH4
Long_RSS	−87 dBm	−87 dBm	−87 dBm
RSS_ACC_MIN	−101 dBm	−105 dBm	−109 dBm
MS_ACC_PWR − P	−8	0	0
max(MS_ACC_PWR − P, 0)	0	0	0
C_RESEL	−87 + 101= 14	−87 + 105 = 18	−87 + 109 = 22
Eligibility	Eligible	Eligible	Eligible

current DCCH, and the mobile needs to leave it. Word errors can occur even in high signal strength conditions if the interfering signal is also strong.

Assuming that during CAND_1 Examination the mobile can synchronize to DCCH4, read the F-BCCH, and confirm that it is the correct neighbor and is not barred, the mobile marks DCCH4 as its current DCCH during Suitable CAND_1 Found and enters the DCCH Camping state. The word error counter is reset and the mobile is hopefully a happy camper again.

9.6.2 Cell Barred

The mobile is camping happily on DCCH1 and monitoring its PCH when it reads a layer 2 go away flag set to 1. The mobile declares a Cell Barred condition and immediately invokes the Candidate Eligibility Filtering procedure. The CEF2 algorithm is executed with DCCH2, DCCH3, and DCCH4 as candidates. The mobile's received signal strengths (Long_RSS) on the DCCHs are DCCH1 = −85 dBm, DCCH2 = −87 dBm, DCCH3 = −87 dBm, and DCCH4 = −87 dBm. Table 9.7 summarizes the calculations the mobile makes during execution of the CEF2 algorithm.

Table 9.7
CEF2 Calculations for a Cell Barred Example

Parameter	DCCH1	DCCH2	DCCH3	DCCH4
Long_RSS	−85 dBm	−87 dBm	−87 dBm	−87 dBm
RSS_ACC_MIN	−105 dBm	−101 dBm	−105 dBm	−109 dBm
MS_ACC_PWR − P	0	−8	0	0
max(MS_ACC_PWR − P, 0)	0	0	0	0
C_RESEL	−85 + 105 = 20	−87 + 101 = 14	−87 + 105 = 18	−87 + 109 = 22
CELLTYPE	—	Preferred	Regular	Nonpreferred
SS_SUFF	—	−89 dBm	—	—
RESEL_OFFSET	—	—	−4 dB	−6 dB
C_RES2	—	—	18 − 20 − 4 = −6	22 − 20 − 6 = −4
Eligibility	—	Eligible (criteria 1)	Ineligible	Ineligible

The mobile enters the Candidate Reselection Rules procedure with only DCCH2 marked as eligible for reselection. DCCH2 is eligible because its CELLTYPE is Preferred and its Long_RSS is greater than its SS_SUFF (−87 dBm −89 dBm). DCCH3 and DCCH4 are ineligible because their C_RES2 are less than zero. This is primarily due to the fact that the Long_RSS on DCCH1 is higher than the Long_RSS on these neighbors, which are not Preferred neighbors. The mobile does not have much to do in the CAND_1 Determination subprocedure because there is only one eligible candidate. If the Long_RSS on DCCH2 had been 2 dB lower, there would not have been any eligible candidates. If this had occurred, the mobile would still have chosen DCCH2 as CAND_1 in the CAND_1 Determination subprocedure because its C_RESEL is greater than 0 and its CELLTYPE is Preferred.

Assuming that during CAND_1 Examination the mobile can synchronize to DCCH2, read the F-BCCH, and confirm that it is the correct neighbor and is not barred, the mobile marks DCCH2 as its current

DCCH during Suitable CAND_1 Found and enters the DCCH Camping state.

9.6.3 Server Degradation

The mobile is camping happily on DCCH1 when the mobile user turns a corner and the signal strength drops rapidly below -105 dBm (RSS_ACC_MIN). This causes C_RESEL to fall below zero, and Short_RSS to fall below Long_RSS. When this occurs, the mobile declares a Server Degradation condition and immediately invokes the Candidate Eligibility Filtering procedure. The CEF2 algorithm is executed with DCCH2, DCCH3, and DCCH4 as candidates. The mobile's received signal strengths (Long_RSS) on the DCCHs are DCCH1 = -107 dBm, DCCH2 = -91 dBm, DCCH3 = -85 dBm, DCCH4 = -101 dBm. Table 9.8 summarizes the calculations the mobile makes during execution of the CEF2 algorithm.

Table 9.8
CEF2 Calculations for a Server Degradation Example

Parameter	DCCH1	DCCH2	DCCH3	DCCH4
Long_RSS	−107 dBm	−91 dBm	−85 dBm	−101 dBm
RSS_ACC_MIN	−105 dBm	−101 dBm	−105 dBm	−109 dBm
MS_ACC_PWR − P	0	−8	0	0
max(MS_ACC_PWR − P, 0)	0	0	0	0
C_RESEL	−107 + 105 = −2	−91 + 101 = 10	−85 + 105 = 20	−101 + 109 = 8
CELLTYPE	—	Preferred	Regular	Nonpreferred
SS_SUFF	−89 dBm	−89 dBm	—	—
RESEL_OFFSET	—	−2 dB	−4 dB	−6 dB
C_RES2	—	10 + 2 − 2 = 10	20 + 2 − 4 = 18	8 + 2 − 6 = 4
Eligibility	—	Eligible (criteria 2)	Eligible (criteria 2)	Eligible (criteria 2)

The mobile enters the Candidate Reselection Rules procedure with all three candidates marked as eligible for reselection. Because all three are eligible as a result of criteria 2, the candidate with the maximum C_RES2 is selected as CAND_1 in the CAND_1 Determination subprocedure. This means that DCCH3 is selected as CAND_1.

Assuming that during CAND_1 Examination the mobile can synchronize to DCCH3, read the F-BCCH, and confirm that it is the correct neighbor and is not barred, the mobile marks DCCH3 as its current DCCH during Suitable CAND_1 Found and enters the DCCH Camping state.

9.6.4 Directed Retry

The mobile is camping on DCCH1 when the user decides to originate a call. The mobile formulates an Origination message, sends it to the network, and enters the Origination Proceeding state waiting for a response from the network. Due to a large volume of traffic in the sector covered by DCCH1, there are no DTCs or AVCs available to assign to the mobile. The network sends a Directed Retry message to the mobile, and the mobile immediately invokes the Candidate Eligibility Filtering procedure. The CEF3 algorithm is executed with DCCH2 and DCCH3 as candidates. DCCH4 is not included as a candidate because it is not listed as a directed retry channel in DCCH1's neighbor list. The mobile's received signal strengths (Long_RSS) on the DCCHs are DCCH1 = −85 dBm, DCCH2 = −87 dBm, and DCCH3 = −87 dBm. Table 9.9 summarizes calculations the mobile makes during execution of the CEF3 algorithm.

The mobile enters the Candidate Reselection rules procedure with DCCH2 and DCCH3 marked as eligible for reselection based on criteria 2. While the calculation of C_RES2 is not necessary to determine the eligibility of the candidates under CEF3, it is necessary for CAND_1 Determination. The candidate with the maximum C_RES2 (DCCH3) becomes CAND_1. Note that even though C_RES2 is negative, DCCH3 still becomes CAND_1 because its C_RES2 is greater than that of DCCH2.

Assuming that during CAND_1 Examination the mobile can synchronize to DCCH3, read the F-BCCH, and confirm that it is the correct neighbor, the mobile selects DCCH3 and returns to the Origination

Table 9.9
CEF3 Calculations for a Directed Retry Example

Parameter	DCCH1	DCCH2	DCCH3
Long_RSS	−85 dBm	−87 dBm	−87 dBm
RSS_ACC_MIN	−105 dBm	−101 dBm	−105 dBm
MS_ACC_PWR − P	0	−8	0
max(MS_ACC_PWR − P, 0)	0	0	0
C_RESEL	−85 + 105 = 20	−87 + 101 = 14	−87 + 105 = 18
CELLTYPE	—	Preferred	Regular
SS_SUFF	−89 dBm	−89 dBm	−99 dBm
RESEL_OFFSET	—	−2 dB	−4 dB
C_RES2	—	14 − 20 − 2 = −8	18 − 20 − 4 = −6
Eligibility	—	Eligible (criteria 2)	Eligible (criteria 2)

Proceeding state. The mobile formulates a new Origination message and sends it to the network on DCCH3. If the network has the resources to set up the call on the sector served by DCCH3, it respond to the mobile with a designation message to a DTC or AVC.

9.6.5 Priority System

For this example, it is assumed that DCCH2 is not on the neighbor list of DCCH1. Instead, DCCH2 serves an autonomous system and broadcasts a PSID that the mobile has stored in its memory (see Chapter 14). Along with the PSID, the mobile has stored a PSP that contains DCCH1 and a POF that corresponds to the channel containing DCCH2.

When the mobile enters the DCCH Camping state on DCCH1, it detects a PSP match for the PSID stored in its memory. It declares a Priority System condition and invokes the Candidate Eligibility Filtering procedure. Because SERV_SS on DCCH1 is greater than zero, the mobile executes CEF4. Besides DCCH2, the mobile evaluates DCCH3 and DCCH4 because they are on DCCH1's neighbor list. The mobile's received signal strengths (Long_RSS) on the DCCHs are DCCH1 = −81 dBm, DCCH2 = −87 dBm, DCCH3 = −83 dBm, and DCCH4 = −111 dBm. Table 9.10 summarizes the calculations the mobile makes during execution of the CEF4 algorithm.

The mobile enters the Candidate Reselection Rules procedure with both DCCH2 and DCCH3 marked as eligible for reselection. DCCH2 is marked as CAND_1 in the CAND_1 Determination subprocedure, however, because it was marked as eligible based on criteria 2 (it has a CELLTYPE of Preferred), while DCCH3 was marked as eligible based on criteria 7 (it has a Regular CELLTYPE). Even though DCCH3 has a

Table 9.10
CEF4 Calculations for a Priority System Example

Parameter	DCCH1	DCCH2	DCCH3	DCCH4
Long_RSS	−81 dBm	−87 dBm	−81 dBm	−111 dBm
RSS_ACC_MIN	−105 dBm	−101 dBm	−105 dBm	−109 dBm
MS_ACC_PWR − P	0	−8	0	0
max(MS_ACC_PWR − P, 0)	0	0	0	0
C_RESEL	−81 + 105 = 24	−87 + 101 = 14	−81 + 105 = 24	−111 + 109 = −2
CELLTYPE	—	Preferred	Regular	—
SS_SUFF	−89 dBm	−89 dBm	—	—
SERV_SS	6	—	—	—
SS_SUFF + SERV_SS	—	−89 + 6 = −83	—	—
RESEL_OFFSET	—	−2 dB	−4 dB	—
C_RES2	—	14 − 24 − 2 = −12	24 − 24 − 4 = −4	—
SS_SUFF − SERV_SS	—	−89 − 6 = -95	—	—
Eligibility	—	Eligible (criteria 2)	Eligible (criteria 7)	Ineligible

higher Long_RSS than DCCH2, the Reselection procedure guides the mobile to the control channel serving the higher priority cell.

Assuming that during CAND_1 Examination the mobile can synchronize to DCCH2, read the F-BCCH, and confirm that it is the correct neighbor and is not barred, the mobile marks DCCH2 as its current DCCH during Suitable CAND_1 Found and enters the DCCH Camping state. The mobile has seamlessly obtained service on the autonomous system without relying on neighbor list information from DCCH1.

9.6.6 Periodic Evaluation

The mobile is camping on DCCH1 and detects a Periodic Evaluation condition every SCANINTERVAL, then invokes the candidate eligibility filtering procedure. The CEF5 algorithm is executed with DCCH2, DCCH3, and DCCH4 as candidates. At this particular time, the mobile's received signal strengths (Long_RSS) on the DCCHs are DCCH1 = −91 dBm, DCCH2 = −101 dBm, DCCH3 = −83 dBm, and DCCH4 = −87 dBm. Table 9.11 summarizes the calculations the mobile makes during execution of the CEF5 algorithm.

The mobile enters the Candidate Reselection Rules procedure with DCCH3 and DCCH4 marked as eligible for reselection. DCCH2 is ineligible because its C_RESEL is not greater than zero. Because DCCH3 has the higher C_RES2 of the two eligible candidates, it is marked as CAND_1 in the CAND_1 Determination subprocedure. Assuming that during CAND_1 Examination the mobile can synchronize to DCCH3, read the F-BCCH, and confirm that it is the correct neighbor and is not barred, the mobile marks DCCH3 as its current DCCH during Suitable CAND_1 Found and enters the DCCH Camping state.

9.7 Implementing hierarchical cell structures

HCS can be used to cause a mobile to reselect to a DCCH on a desired cell or sector before it would otherwise do so, then stay connected to it for as long as possible. This is an attractive capability for the service provider that allows him or her to optimally design the network. Taking full advantage of HCS requires an understanding of the nuances of reselection

Table 9.11
CEF5 Calculations for a Periodic Evaluation Example

Parameter	DCCH1	DCCH2	DCCH3	DCCH4
Long_RSS	−91 dBm	−101 dBm	−83 dBm	−87 dBm
RSS_ACC_MIN	−105 dBm	−101 dBm	−105 dBm	−109 dBm
MS_ACC_PWR − P	0	−8	0	0
max(MS_ACC_PWR − P, 0)	0	0	0	0
C_RESEL	−91 + 105 = 14	−101 + 101 = 0	−83 + 105 = 22	−87 + 109 = 22
CELLTYPE	—	—	Regular	Nonpreferred
SS_SUFF	−89 dBm	—	—	—
SERV_SS	6	—	—	—
SS_SUFF − SERV_SS	−89 − 6 = −95	—	—	—
RESEL_OFFSET	—	—	−4 dB	−6 dB
C_RES2	—	—	22 −14 − 4 = 4	22 − 14 − 6 = 2
Eligibility	—	Ineligible	Eligible (criteria 7)	Eligible (criteria 6)

described in the preceding sections. However, the following general rules can be applied to the implementation of HCS:

- RSS_ACC_MIN and MS_ACC_PWR are used to define the coverage area of the cell or sector served by a DCCH.

- DELAY is used to keep fast-moving mobiles from reselecting to DCCHs of microcells or picocells.

- CELLTYPE is used to cause a mobile to prefer one DCCH over another when the received signal strength on both DCCHs is sufficient for service.

- SS_SUFF is used to ensure a mobile does not reselect to a DCCH before the received signal strength on that DCCH is above an absolute threshold and to keep the mobile connected to a DCCH until the signal strength on that DCCH falls below the absolute threshold.

- SERV_SS is used as a bias to keep a mobile from reselecting from a DCCH.

- RESEL_OFFSET is used to ensure a mobile does not reselect to another DCCH until the received signal strength on the second DCCH is sufficiently above that of the current DCCH.

The best way to see these general rules is to examine the CEF5 algorithm, flowcharted in Figure 9.5, which is applied every time the mobile periodically evaluates neighboring control channels. The first check the mobile makes is to ensure it is within the coverage area of the candidate—that is, that C_RESELcand is greater than zero as defined by RSS_ACC_MINcand and MS_ACC_PWRcand. The next check is to ensure the mobile has been within the coverage area of the candidate for the specified amount of time, DELAYcand. The third check is to identify the CELLTYPE of the candidate, which in turn determines the remainder of the filtering that is applied to the candidate and results in a prioritization of Preferred candidates over Regular ones, and Regular candidates over Nonpreferred ones. The next check for a Preferred candidate is to determine if the received signal strength is above the absolute thresholds for the candidate, SS_SUFFcand and SERV_SScand. For a Nonpreferred candidate, the check is instead to determine if the received signal strength on the current DCCH is below its absolute thresholds, SS_SUFFcur and SERV_SScur. Finally, if the candidate is not Preferred, the mobile checks to ensure the received signal strength of the candidate is above a relative threshold, RESEL_OFFSET, with respect to the current DCCH.

An HCS example will help explain these rules more fully. Suppose a macrocell is covering a busy intersection of two highways. The macrocell is experiencing congestion due to the large volume of traffic in this area. The service provider decides to locate a microcell close to the intersection to gain capacity and offload traffic from the macrocell. This situation is depicted in Figure 9.6.

A good way to implement HCS for this example would be to make the microcell a Preferred neighbor to the macrocell, and the macrocell a

168 Understanding Digital PCS

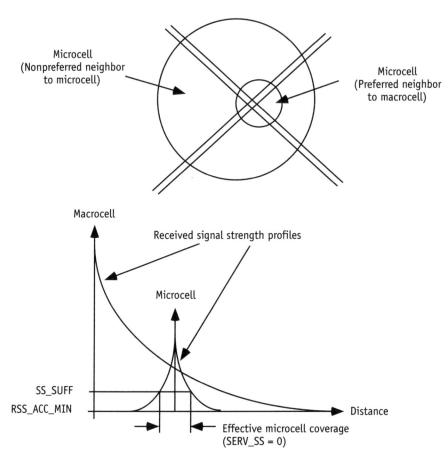

Figure 9.6 An example of HCS implementation.

Nonpreferred neighbor to the microcell. SS_SUFF for the microcell would be set to the absolute signal strength level at which reselection to the microcell should occur. This allows reselection into the microcell at a received signal strength level that could be below the received signal strength level of the macrocell. It also allows the mobile to stay camping on the microcell's DCCH well beyond the point where reselection would occur if the macrocell was treated as a Regular neighbor to the microcell. The macrocell's neighbor list entry for the microcell's DCCH should also include a DELAY adequate to keep fast-moving traffic from reselecting to the microcell. It would not be optimal if fast-moving mobiles accessed the microcell and attempted to set up a call that would require an immediate handoff to another cell.

Reference

[1] Telecommunications Industry Association, *TIA/EIA Interim Standard: TDMA Cellular/PCS—Radio Interface—Mobile Station-Base Station Compatibility—Digital Control Channel, TIA/EIA/IS-136.1-A*, Oct. 1996, Section 6.3.3.

10

Mobile Sleep Mode

INCREASED STANDBY time is one of the advantages of Digital PCS over AMPS. It is made possible primarily by the implementation of *sleep mode* in mobiles. Sleep mode is the state in which a mobile switches off devices for a period of time when they are not needed. This reduces the current drain from the battery and increases the time before the battery needs to be recharged. The length of time that a mobile can be powered on and available for service on a control channel before the mobile's battery must be recharged is called the *standby time*. The standby time of a Digital PCS mobile can be greater than that of an AMPS mobile because operation on the DCCH allows the mobile to enter sleep mode for relatively long periods of time. Standby time can be a marketing advantage of one mobile over another. Not being able to place or receive a call, or dropping a call in progress, due to a dead battery can be a major annoyance to a mobile user. This chapter explores the capabilities available to a mobile while operating on a DCCH to take advantage of sleep mode and increase standby time.

10.1 Paging frame class and PCH subchannels

The primary DCCH characteristic that allows a mobile to take advantage of sleep mode is the division of the SPACH into paging frames and PCH subchannels [1]. Each mobile is only required to periodically monitor a small number of time slots to determine if the network desires to send it information. The period between monitoring these time slots is determined by the paging frame class (PFC) at which the mobile is operating. The primary time slot that the mobile periodically monitors is called its assigned PCH subchannel. During the periods when the mobile is not monitoring time slots for network-originated messages or performing other processing, it may shut down or significantly reduce the current drain of its receiver and logic circuitry.

The PFC determines the periodicity in hyperframes of the PCH subchannel that a mobile is required to monitor. As shown in Table 10.1, eight PFCs have been defined for Digital PCS. Not all of these PFCs are applicable to traditional voice service, however. Excessive delay is introduced when paging a mobile operating at a PFC greater than 3 while a caller is waiting at the other end. PFCs greater than 3 may be used when a mobile is not in a mode to receive voice calls, such as in data-only mode.

Table 10.1
The Eight Paging Frame Classes

PFC	Periodicity of PCH (in hyperframes)	Periodicity of PCH (in seconds)
1	1	1.28
2	2	2.56
3	3	3.84
4	6	7.68
5	12	15.36
6	24	30.72
7	48	61.44
8	96	122.88

For any given PFC, a mobile's PCH subchannel occurs twice—once in the primary superframe and once in the secondary superframe. This is illustrated in Figure 10.1 for PFCs 1, 2, and 3. If a mobile misses reading its PCH subchannel in the primary superframe, it may still be able to read it in the secondary superframe. This is important for two reasons. First, the mobile may not be able to synchronize to the PCH subchannel in the primary superframe due to poor signal strength conditions, interference, or other channel impairments. Second, the mobile may be performing other tasks at the time the PCH subchannel occurs on the primary superframe, and may not be able to tune its receiver or process the received signal. In these cases, the mobile may still examine the PCH subchannel on the secondary superframe to determine if a message is present and addressed to it.

Most first-generation Digital PCS networks and mobiles only operate at PFC1. To operate at a higher PFC, both the network and mobile must support the desired PFC. This is a more difficult task for the network than the mobile, as the network must buffer and correctly time the transmission of messages to mobiles operating at different PFCs. The Digital PCS standard has been designed to allow the network to dynamically change the PFC at which it allows mobiles to operate. This can be useful to

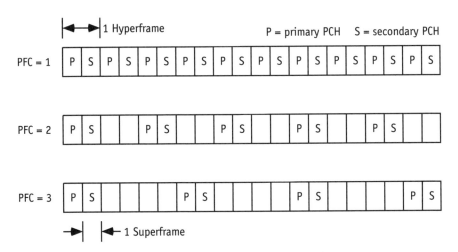

Figure 10.1 PCH subchannels on the primary and secondary superframes for PFCs 1, 2, and 3.

increase the capacity of the DCCH in heavy usage times, because higher PFCs allow the network to page more mobiles. The network broadcasts the value of the highest PFC it supports in the MAX_SUPPORTED_PFC information element in the DCCH Structure message on the F-BCCH.

When a mobile camps on a DCCH, it uses PFC1 until it has registered. The network may inform the mobile of the PFC it is to use by sending the PFC Assignment information element in the Registration Accept message sent to the mobile on the SPACH. Then, as the mobile monitors its PCH subchannel, the network may indicate a change to the assigned PFC by setting the paging frame modifier (PFM) bit to 1 in the layer 2 SPACH header A. If PFM is set to 1, the mobile must either push out or pull in its current PFC by 1. If the PFM_DIRECTION information element in the DCCH Structure message is set to 1, the mobile must push out its PFC by 1. If PFM_DIRECTION is set to 0 instead, the mobile must pull in its PFC by 1. This allows the network to dynamically change the current PFC at which the mobile is operating.

Changing the PFC at which a mobile is operating impacts the mobile's sleep mode. In general, increasing the PFC increases the length of time that the mobile may stay in sleep mode, thereby increasing standby time. Conversely, decreasing the PFC decreases the standby time because the mobile cannot stay in sleep mode as long as it would at a higher PFC. However, this is only true up to a point. There is a law of diminishing returns as the PFC is increased, and there is a point at which the current required to turn devices back on is greater than the current required to keep them on.

The network may require a mobile to monitor more time slots than its assigned PCH subchannel. This may be done because the network has a message for the mobile but cannot send it in the mobile's assigned PCH subchannel due to network load. The process of instructing a mobile to read additional time slots after its assigned PCH subchannel is called PCH Displacement. When a mobile reads its assigned PCH subchannel and does not find a message addressed to it, it must examine the PCON bit sent at layer 2 in the SPACH Header A of that time slot. If the PCON bit is set to 0, the mobile may sleep until its next assigned PCH subchannel. If the PCON bit is set to 1, however, the mobile must continue to read SPACH slots for a message directed to it. The number of time slots the mobile must read is determined by the PCH_DISPLACEMENT information element sent in the DCCH Structure message. The mobile must read until it receives a message addressed to it, or it reads the number of time slots indicated by PCH_DISPLACEMENT.

The PCH Displacement process can impact a mobile's sleep mode if it occurs often. The more time slots the mobile must read, the less time the mobile has to stay in sleep mode. The most efficient operation from the perspective of maximizing the time the mobile can stay in sleep mode is to keep PCON set to 0 at all times. Unfortunately, this may not be the most efficient use of network resources, particularly if the mobile is operating at a higher PFC. Without PCH Displacement, the network may be required to buffer a message destined for a mobile until the mobile's next PCH subchannel arrives. This increases buffer size requirements and message delivery delay.

10.2 Processes impacting sleep mode

A Digital PCS mobile is required to perform processing in addition to monitoring its assigned PCH subchannel. This additional processing impacts sleep mode because the mobile cannot power down various components as much as it would otherwise. The major processes that impact sleep mode include neighbor channel measurements, Mobile Assisted Channel Allocation (MACA) measurements, and rereading the BCCH [2]. The impact of each of these processes on mobile sleep mode is discussed in this section.

10.2.1 Neighbor channel measurements

Neighbor channel measurements are made periodically as part of the Reselection procedure. These channel measurements require the mobile to tune its receiver to different channels and measure signal strength, then process these measurements to determine if reselection to a different control channel is required. Neighbor channel measurements must be performed during the period between monitoring the mobile's PCH subchannel. The number of neighbors that must be measured and the interval between measurements is defined by information broadcast on the

DCCH. The more neighbor channels that must be measured and the more frequently they must be measured, the less time the mobile has to stay in sleep mode.

The number of neighbor channels requiring measurement is identified in the neighbor cell message on the E-BCCH. There may be up to 24 neighbor channels (both DCCHs and ACCs) listed in the message. The signal strength measurement process is no different for ACCs than for DCCHs. There may be other channels, called private operating frequencies (POFs), that must be measured in addition to the neighbor channels. POFs are described in Chapter 14. They are channels that potentially contain one or more DCCHs for a nonpublic system. There are typically four POFs for each nonpublic system. The signal strength on the current DCCH must also be measured for comparison purposes with the signal strength of the neighbor channels and POFs. It may be possible for the mobile to measure the signal strength on the current DCCH at the same time it monitors its PCH subchannel, however.

The measurement intervals for the neighbor channels, POFs, and current DCCH may be different, and are defined by the SCANINTERVAL and HL_FREQ parameters broadcast on the current DCCH. SCANINTERVAL is an information element sent in the Control Channe l Selection Parameters message on the F-BCCH. It identifies the basic interval in hyperframes between signal strength measurements. SCANINTERVAL can be varied from 1 to 16 hyperframes. HL_FREQ is a field sent for each neighbor channel in the Neighbor Cell or Neighbor Cell (Multihyperband) message on the E-BCCH. It modifies the measurement interval for the neighbor channel it is associated with. If HL_FREQ is set to 1, the measurement interval is SCANINTERVAL. If HL_FREQ is set to 0, the measurement interval is twice SCANINTERVAL. The measurement interval for the current DCCH and all POFs is always SCANINTERVAL. An example of the different measurement intervals for the current DCCH, neighbor channels, and POFs is provided in Figure 10.2.

SCANINTERVAL and HL_FREQ are parameters under service provider control. To maximize mobile standby time, SCANINTERVAL should be set large and HL_FREQ should be set to 0. However, these are typically not appropriate settings because timely reselection to the best server is important to optimize network performance. The longer the

Mobile Sleep Mode   177

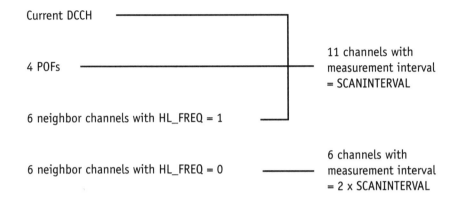

Number of channels requiring measurement every SCANINTERVAL:

11 + 6/2 = 14 channels

Figure 10.2   Example of measurement intervals for reselection purposes.

measurement interval, the longer it takes a mobile to reselect to a better server. Long measurement intervals therefore cause dragged reselections. This is particularly true for macrocells intending to carry vehicular traffic. Longer SCANINTERVALs are most appropriate for cells in which mobiles are not expected to move very rapidly.

The processing associated with the neighbor channels is described in Chapter 9. The signal strength measurements are averaged, and at least once every SCANINTERVAL the Candidate Eligibility Filtering procedure is applied to all the channels for which an adequate number of signal strength measurements have been made. The Candidate Reselection Rules are applied to the eligible channels, and the best candidate for reselection is chosen. The best candidate is examined to determine if it is suitable for reselection. This examination requires the mobile to retune to the channel, synchronize to it, and read a portion of the F-BCCH. All of this processing reduces the time the mobile may spend in sleep mode, thereby reducing the standby time. Reselection is a necessary process for ensuring the mobile is operating on the optimum control channel, however.

10.2.2 Mobile Assisted Channel Allocation

MACA measurements are made by a mobile if the network instructs the mobile to do so. The network may use these measurements for adaptive channel allocation purposes (see Chapter 2). The type of measurements made for MACA channels are similar to those made for neighbor channels. The mobile must tune its receiver to the MACA channels one at a time and perform signal strength measurements on each of them. When instructed to do so by the network, the mobile must report these measurements to the network in a MACA Report message.

The network may broadcast a list of MACA channels in the Mobile Assisted Channel Allocation message on either the F-BCCH or E-BCCH. This list may be up to 15 channels in length. The network may request the mobile to send a MACA Report in response to an audit, in response to a page, when the mobile is registering, or when the mobile is originating. A mobile may be allowed to only report the signal strength on the first eight channels in the MACA list in response to a page or when originating. The specific type of MACA that requires the mobile to make signal strength measurements on channels other than the current DCCH is known as short-term MACA (STM). A mobile may also be required to report what is called long-term MACA (LTM), which consists of measurements of the signal strength, word error rate (WER), and bit error rate (BER) on the current DCCH. All of these parameters that control MACA reporting are included in the Mobile Assisted Channel Allocation message.

There are well-defined rules that govern the STM and LTM measurement and processing techniques. As LTM measurements and processing can be performed by the mobile as a part of monitoring the mobile's PCH subchannel, only STM measurements and processing will be considered here. A minimum of four signal strength measurements must be made and averaged on every MACA channel prior to reporting STM to the network. The minimum time between two consecutive signal strength measurements on the same frequency is 20 ms, and the maximum time is 500 ms.

With the STM measurement and processing requirements just described, it is possible for a mobile to only make STM measurements on MACA channels immediately before they must be reported to the network. For example, a mobile has a time window of one superframe (640 ms) in which to respond to an audit or page from the network. If it took the mobile 5 ms to tune to a channel and make a signal strength measurement, then 15 channels could be measured in $15 \times 5 = 75$ ms. Repeating these measurements four times for averaging purposes would require $75 \times 4 = 300$ ms, which is well within the time window required to respond to the audit or page. Such "just-in-time" measurements keep the mobile from continuously measuring MACA channels, which could significantly reduce the time available for sleep mode.

If MACA is used in support of adaptive channel allocation, the best way the service provider can ensure that mobile standby time is not significantly impacted by MACA reporting is to minimize the number of channels in the MACA list. This gives the mobile more time in sleep mode. If required, the Mobile Assisted Channel Allocation message can be changed to add, delete, or replace channels in the MACA list. Mobiles that are camping on the DCCH are notified of a change to the broadcast control channel as described in the next section.

10.2.3 Rereading the broadcast control channel

Another mobile process that can limit the benefits of sleep mode is frequent rereading of the BCCH due to updates. When critical information is updated on either the F-BCCH or E-BCCH, all mobiles camping on the DCCH are notified that information has changed. The mobiles are required to reread all or part of the BCCH to identify what information has changed. Changes to critical BCCH information should be minimized to limit the impact on mobile sleep mode.

A mobile monitoring its assigned PCH subchannel, or receiving a message on the SMSCH or ARCH, is notified of a change to the BCCH when the broadcast change notification (BCN) bit is toggled. The BCN flag is provided in the layer 2 SPACH Header A field. If the mobile has not received a message intended for it on its PCH subchannel, then during the next hyperframe the mobile must tune to the F-BCCH and read the F-BCCH change (FC) bit and the E-BCCH change (EC) bit at layer 2. These bits toggle to indicate a change to their respective contents. The mobile is required to reread a full cycle of either or both of these BCCHs if a change is indicated. As the E-BCCH may extend over many superframes, this could require the mobile to remain active for a relatively long period of time to ensure the entire cycle is read. Because the toggling of the BCN, FC, and EC bits is used to indicate a change to BCCH contents, the network should not change the contents more often than a mobile that is operating at the maximum supported PFC can handle. If the BCCH is changed too frequently, the change notification bits may toggle an even number of times before the mobile monitors its assigned PCH subchannel. If this occurs, the mobile will be unaware that a change to the BCCH has occurred.

Information broadcast on the BCCH is identified as noncritical in the standard if changes to it do not require that BCN be toggled. An example of this type of information is the Time and Date message, which can be broadcast on the E-BCCH. The network may, however, toggle the BCN when noncritical information changes if it so desires. Critical BCCH changes are typically rare. An example of critical changes to the BCCH occurs when the service provider is updating network parameters to optimize performance, such as updating the contents of the Neighbor Cell message when a new cell site is installed. The example was given in the previous section of changing the MACA list to support adaptive channel allocation. If the MACA list is updated in this manner, it may be desirable to place the Mobile Assisted Channel Allocation message on the F-BCCH instead of the E-BCCH, so mobiles are not required to reread a full cycle of the E-BCCH when a MACA list change occurs.

Although the S-BCCH is not supported by first-generation Digital PCS mobiles and networks, it will be in the future. Changes to the S-BCCH could impact mobile sleep mode as broadcast teleservices are standardized, developed, and deployed. Toggles similar to the BCN, FC, and EC are used with the S-BCCH to identify a change to the broadcast contents. Overhead information on one of the frequently transmitted S-BCCH subchannels can be used to identify exact changes to the S-BCCH. This allows mobiles to quickly determine if information of interest to it has changed. If the mobile is not interested in the information, it may return to sleep mode. Changes to the S-BCCH can also be broadcast on one of the frequently transmitted S-BCCH subchannels to allow mobiles to quickly read the new information and return to sleep mode. Additionally, it is always optional for a mobile to read information on the S-BCCH, so a mobile that is attempting to maximize standby time is not required to read the S-BCCH if a change to its contents occurs.

10.3 Extending sleep mode with optional enhancements

A Digital PCS mobile may implement optional enhancements to neighbor channel measurements in order to extend sleep mode. If a mobile implements these enhancements and the network indicates they may be used (by setting the scanning option indicator to 1 in the Control Channel Selection Parameters message on the DCCH), the measurement interval on one or more neighbor channels may be increased to extend sleep mode. The conditions under which the measurement interval may be increased require that the RF environment has remained relatively static for some period of time. Such static conditions are most likely to occur when a mobile is stationary.

There are three conditions under which the measurement interval on neighbor channels may be increased. The first is if the time since the last reselection is greater than one hour. If this is the case, the mobile is allowed to increase the measurement interval for all neighbor channels by a factor of two. If another hour passes without a reselection, the measurement interval may once again be increased by a factor of two. The measurement interval is returned to its value as defined by SCANINTERVAL and HL_FREQ if a reselection does occur.

The second condition for increasing the measurement interval requires the mobile to average the last 25 signal strength measurements on the current DCCH and the neighbor channels. This averaged signal strength measurement is called the processed signal strength (PSS). If the PSS on the current DCCH and the PSS on every neighbor channel does not change by 7 dB or more in 5 minutes, then the mobile is allowed to increase the measurement interval by a factor of two for every neighbor channel. The increased measurement interval is additive, and it must be returned to the original measurement interval when the condition is no longer valid.

The third condition for activating an optional enhancement allows the measurement interval to be increased for individual neighbor channels instead of all the channels. If the difference in PSS between the current DCCH and a neighbor channel has not changed by 10 dB or more for the last 5 minutes, the mobile is allowed to increase the measurement interval for the neighbor channel by a factor of two. Once again, the measurement interval extension is additive and must be removed when the condition is no longer valid.

These optional enhancements can extend sleep mode and increase standby time because they may be applied to neighbor channel measurements. There are other ways to increase sleep mode, and to increase the effectiveness of sleep mode on standby time. Other methods include using lower voltage components, turning off more components when they are not needed, and performing measurements and processing in clever ways (for example, only performing MACA measurements immediately prior to an access requiring a MACA report).

References

[1] Telecommunications Industry Association, *TIA/EIA Interim Standard: TDMA Cellular/PCS—Radio Interface—Mobile Station-Base Station Compatibility—Digital Control Channel, TIA/EIA/IS-136.1-A*, Oct. 1996, Section 4.

[2] Ibid., Section 6.

Voice Services

DIGITAL PCS SUPPORTS voice services on both the AVC and the DTC. Analog voice processing is well known, as are the call-processing procedures that support it in AMPS and Digital PCS. This chapter focuses on digital voice service provided on a DTC. Advantages of digital voice service over analog include greater privacy, less static, and an increase in talk time resulting from the speech compression process.

11.1 Voice coding and decoding

A speech signal must be digitized and compressed for transmission across a digital channel. This is accomplished in Digital PCS with a voice coder (vocoder). A vocoder is best described as an algorithm that synthesizes speech by creating a model of the speech-generation process. The model consists of an input signal and a speech synthesizer. The speech-generation model is optimized at the transmitting end by minimizing the error between the original and synthesized speech signals in a technique called analysis-by-synthesis [1]. The parameters that define the speech-generation model are transmitted across the digital channel. The vocoder at the receiving end synthesizes the speech by feeding the input signal into the speech synthesizer. This process is shown in Figure 11.1.

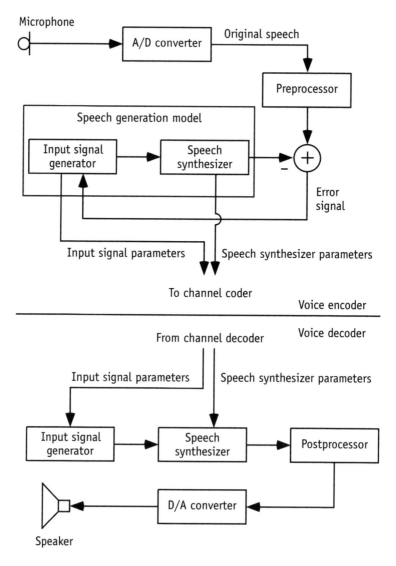

Figure 11.1 Basic operation of a vocoder.

Figure 11.1 also shows the analog-to-digital (A/D) conversion and preprocessing that must occur prior to voice encoding, and the postprocessing and digital-to-analog (D/A) conversion that must occur after voice decoding. The preprocessing typically consists of (1) bandpass filtering to ensure the digitized signal from the microphone only contains components in the voice band, and (2) level adjustment to ensure the digitized signal does not cause an overflow condition in the vocoder algorithm. The preprocessing might also include a noise suppression algorithm to attenuate nonspeech signals because the vocoder is optimized to work with speech. The postprocessing typically consists of bandpass filtering and level adjustment prior to D/A conversion.

Most vocoders use an input signal and speech synthesizer based on the way humans speak [2–5]. Speech is formed through an excitation of the vocal tract, which is composed of the oral cavity, tongue, lips, and nasal cavity. The excitation is formed by air flowing from the lungs through the vocal tract in either a periodic or random fashion, based on the type of sound being made. Sounds may be either voiced or unvoiced. Voiced sounds, like "A," are generated when pulses of air flow through the vocal tract in a nearly periodic fashion. The spacing between pulses of air is called the pitch period. Unvoiced sounds, like "S," are generated by turbulent, or random, air flow through a constriction in the vocal tract. The characteristics of the vocal tract may change approximately every 20 to 30 ms, and the characteristics of the excitation may change approximately ever 2 to 5 ms. This model has been successfully applied to the design of vocoders. The excitation is the input signal to the speech synthesizer. The speech synthesizer is a time-varying digital filter, or synthesis filter, that approximates the characteristics of the vocal tract.

The power of a vocoder lies in its ability to minimize the number of parameters that must be transmitted to define the excitation and synthesis filter to synthesize speech, thereby resulting in a low bit rate across the air interface while maintaining high-quality speech reproduction. A low bit rate is important because the data transmitted across the air interface must be heavily coded to reduce the probability of experiencing errors in the received data pattern. The fewer data bits required to define the excitation and synthesis filter, the more bits available for channel coding. In the past, vocoders in the bit rate range required for transmission across a typical cellular air interface sounded "mechanized" or "hollow" to many listeners. With advances in digital signal processing technology and new vocoder algorithms that more faithfully reproduce speech, these issues have been eliminated.

11.1.1 IS-641 ACELP

The IS-641 vocoder is used almost exclusively in Digital PCS networks. An older vocoder standardized for use in IS-54 has been replaced by IS-641 in most networks due to the exceptional voice quality provided by the newer vocoder [6]. This discussion will focus exclusively on IS-641.

IS-641 uses code excited linear prediction (CELP) as the speech-generation model [7], which is shown in Figure 11.2. In CELP, the excitation is a sequence from a codebook and the synthesis filter is a linear prediction filter. Linear prediction (LP) techniques efficiently exploit the short-term correlation properties in speech signals. Long-term prediction (LTP) is typically also used in CELP vocoders to predict the pitch of the excitation. This leads to an excitation composed of two elements—an adaptive component that predicts the pitch and a fixed codebook component to model the residual excitation, which is sometimes called the innovation. Determining the optimal excitation is the most difficult part of the CELP algorithm. It is typically achieved by searching the codebook for the excitation that minimizes the perceptually weighted error between the original and synthesized speech signals. The perceptually weighted error signal is constructed by passing the error signal through a perceptual weighting filter that attenuates those frequencies where the error is perceptually less important and amplifying those frequencies where the error is perceptually more important. Typically, the larger the codebook, the more realistic the speech can be synthesized. However, a larger codebook can require more storage and processing memory, and increases the complexity of the codebook search.

IS-641 uses a class of CELP known as algebraic CELP (ACELP) to reduce the effects of a large codebook on storage and search complexity [8,9]. In ACELP, the residual portion of the excitation (innovation) is composed of an algebraic codebook and a shaping matrix. The algebraic codebook can be efficiently searched and need not be stored. The shaping matrix, called an adaptive prefilter in IS-641, is applied to the codebook

Voice Services 187

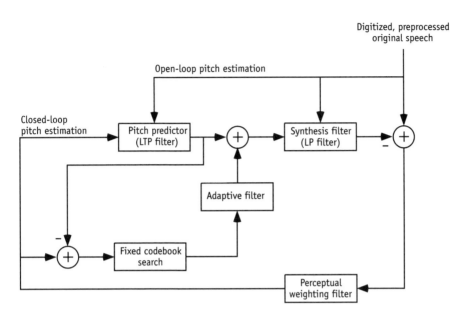

Figure 11.2 The CELP speech-generation model.

and provides greater flexibility in defining the excitation to improve the quality of the synthesized speech.

The IS-641 synthesis filter is realized as a tenth-order linear prediction filter that operates on a sliding 30-ms window of speech to model 20-ms speech frames using the Levinson-Durbin algorithm. This speech windowing is shown in Figure 11.3. In addition to segmenting the speech into 20-ms frames, each frame is divided into 5-ms subframes. The 30-ms window operated on by the linear prediction filter is composed of a 5-ms subframe from the previous speech frame, the current 20-ms speech frame, and a 5-ms subframe called look-ahead from the next speech frame. The linear prediction coefficients for the resulting linear prediction filter are coded into 26 bits per 20-ms speech frame.

The pitch lag and gain of the excitation is predicted through a combination of open-loop and closed-loop analysis. Open-loop pitch analysis is performed twice per speech frame (every 20 ms) and is used to reduce the complexity of the closed-loop pitch analysis, which is performed every subframe. The open-loop pitch analysis is performed by calculating the

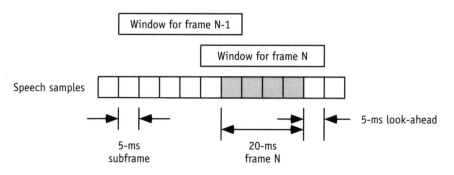

Figure 11.3 Speech windowing used with IS-641.

correlation between a perceptually weighted version of the speech signal and a delayed version, and finding the delay at which the correlation is the greatest. Closed-loop pitch analysis is performed around the open-loop pitch estimate every subframe, and is achieved by minimizing the mean-squared error between the perceptually weighted original and synthesized speech. The pitch lag is encoded with 8 bits in the first and third subframes, and 5 bits in the second and fourth subframes. The pitch gain and the adaptive codebook gain are together coded into 7 bits every subframe.

The algebraic codebook is updated every subframe and is composed of a sequence that contains four nonzero pulses. The pulses can have amplitudes of 1, and there are 40 positions in a subframe where pulses may occur. The algebraic codebook is searched by minimizing the mean-squared error between the perceptually weighted original and synthesized speech. The optimal algebraic codebook is coded as 17 bits per superframe.

Table 11.1 shows the bit allocations for the output of the IS-641 encoder. These bits represent the data that is sent to the channel coder and interleaver for transmission across the radio channel. The encoder yields 148 bits every 20 ms, resulting in a bit rate of 7.4 Kbps. As the usable data rate on a $\pi/4$ DQPSK DTC is 13 Kbps, this leaves 5.6 Kbps for channel coding.

The channel coding used with IS-641 is depicted in Figure 11.4. The 148 data bits per 20 ms from the speech encoder are divided into three classes before channel coding. The three classes are based on the

Table 11.1
IS-641 Encoder Bit Allocations

Parameter	Subframe 1	Subframe 2	Subframe 3	Subframe 4	Total Frame
Linear prediction coefficients	—	—	—	—	26 bits
Pitch lag	8 bits	5 bits	8 bits	5 bits	26
Pitch gain and adaptive codebook gain	7	7	7	7	28
Algebraic codebook	17	17	17	17	68
Total					148 bits

perceptual significance of the bits to the subjective voice quality of the synthesized speech. The more perceptually significant the bits, the more channel coding used to protect them from the harsh RF channel conditions that could cause impairments. There are 48 class 1A bits, 48 class 1B bits, and 52 class 2 bits. A 7-bit CRC code is used with the class 1A bits to ensure they are correctly received at the channel decoder. These are the most perceptually significant bits. If they are in error, a bad frame masking routine is used at the decoder to conceal the errors. The 96 class 1 bits and the CRC for the class 1A bits are protected by a rate 1/2 convolutional code with memory order 5. Five tail bits (set to zero) are added prior to convolutional encoding, so the 96 class 1 bits, 7 CRC bits, and 5 tail bits are encoded as 216 bits. The 216-bit code is punctured by removing 8 bits. The remaining 52 class 2 bits are not protected at all by channel coding. The 208 coded bits and 52 uncoded bits form the 260 bits that are interleaved across two time slots and inserted in the DATA field of the DTC (see Chapter 4).

At the receiving end, the DATA bits are deinterleaved, convolutionally decoded, and sent to the speech decoder where the speech is synthesized as described previously. A bad frame masking routine is used if the

190 Understanding Digital PCS

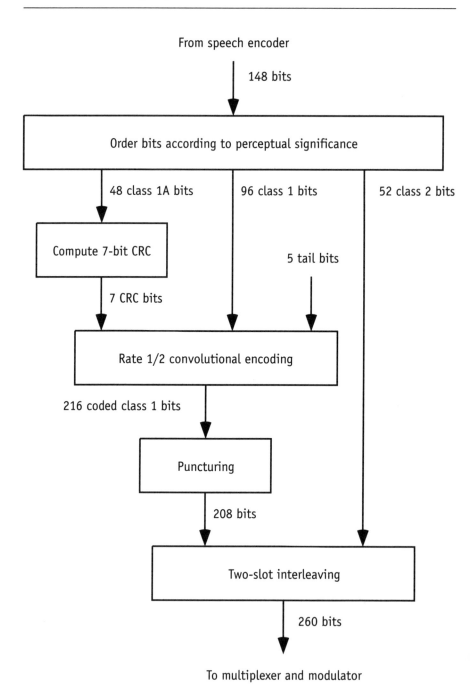

Figure 11.4 Channel coding used with IS-641.

CRC check of the class 1A bits fails. The bad frame masking routine is not standardized for IS-641, although a typical implementation would be to replace the excitation gains for the current speech frame with attenuated versions of the excitation gains from the previous frame. The attenuation is increased if more CRC errors occur in subsequent speech frames, until total muting of the speech occurs. (This is similar to the annoying habit of mumbling a name, any old name, in greeting to someone when you have totally forgotten his or her real name. If you are a good mumbler, the person might not detect that you really do not have a clue what his or her name is.) Of course, the RF channel conditions would have to be very poor for this to occur.

The speech samples provided to the speech encoder are sampled at a rate of 8,000 Hz and uniformly quantized with at least 13 bits of dynamic range. The 8,000-Hz sampling rate fulfills the Nyquist criteria for sampling the speech signal at greater than twice the highest frequency present because human speech ranges from approximately 80 Hz to over 3,000 Hz. The IS-641 speech encoder input expects quantized speech samples represented as 16-bit integers, and the speech decoder outputs synthesized speech samples as 16-bit integers.

11.1.2 Discontinuous transmission and comfort noise

IS-641-A defines the complementary features of discontinuous transmission (DTX) and comfort noise (CN), which are enhancements to the ACELP vocoder. With DTX and CN, voice activity detection is used to classify the signal to be modeled by the vocoder as either speech or background noise [10]. When it is determined that speech is present, the vocoder works as described in the previous section. During silent periods when no voice activity is detected, the vocoder parameters need not be transmitted. Instead, occasional comfort noise parameters are extracted by the speech encoder and transmitted across the air interface. The decoder uses the comfort noise parameters to synthesize the background noise, which is inserted into the received audio to fill the silent periods that would otherwise occur as a result of DTX. Without CN, DTX would cause a complete muting of the output audio during inactive voice segments, resulting in a rapid drop in the signal energy level that is perceptually unpleasant to a listener.

Voice activity detection typically works by extracting parameters from the signal to be modeled by the vocoder, comparing these parameters with background noise parameters, making an initial decision based on the parameter comparison, then smoothing and correcting the decision based on frame energy thresholds and the state of the previous frames. Parameters that may be extracted from the original signal every 20 ms include the linear prediction coefficients from the synthesis filter, the full-band energy in the frame, the low-band (from 0 to 1 kHz) energy in the frame, and the zero-crossing rate. The difference is computed between these parameters and a running average of the same parameters for frames classified as containing only background noise. An initial decision on whether the current frame is composed of voice or background noise is made using the computed difference parameters. The decision is based on a boundary hardcoded into the voice activity detector and set empirically based on known voice and background frames. The decision may be changed in the smoothing step to bias the decision to voice if the previous frames were classified as voice and the frame energy is above a threshold, or to bias the decision towards background noise if the previous frames were classified as background noise and the frame energy is below a threshold.

If it is determined through voice activity detection that only background noise is present during a frame, then comfort noise parameters are calculated by the speech encoder to model the background noise. The background noise model is similar to the model used for speech and may be generated using the same ACELP algorithm. In IS-641-A, only 38 bits are required to model the comfort noise, however, compared to 138 bits to model speech. The comfort noise is generated by the speech decoder at the receiving end, and comfort noise frames are inserted into the voice stream prior to postprocessing. The effects of frame erasures tend to be less severe when DTX/CN is used, because there are fewer voice frames transmitted and the erasure of a comfort noise frame does not cause a severe degradation in voice quality.

Once the comfort noise parameters have been sent to the speech decoder, they only need to be updated when the background noise changes significantly. This allows the transmitter to be shut off during periods of voice inactivity except when comfort noise parameters must be transmitted. Because a significant portion of a conversation is made up of silence—up to 60% [11]—DTX can dramatically increase talk time. It can also reduce the uplink interference, which can lead to higher voice quality. Because many cellular networks measure the mobile's uplink signal strength to determine when to hand off the mobile to another channel, completely powering down the mobile's transmitter during silent periods can impact network performance. A mode of operation has been defined in Digital PCS in which a mobile operating in DTX mode will always transmit a short burst on its assigned time slot so the network can still measure received signal strength on the uplink.

11.1.3 Vocoder implementation

Digital PCS mobiles are required to implement a bit-exact, fixed-point ANSI C code version of IS-641. A bit-exact vocoder implementation allows for simpler testability and voice quality consistency between different mobiles. The fixed-point vocoder implementation is important because most digital signal processors (DSPs) used in mobiles are fixed point, not floating point. Conversely, flexibility is allowed in the implementation of IS-641 in the network because more powerful floating point DSPs are used. Network implementations of IS-641 are therefore not required to be bit-exact, but they are subject to the minimum performance requirements for IS-641, which are contained in IS-686.

Different vocoder placement strategies may be used in the network [12]. Vocoders may either be placed at the base station or at the MSC. Locating the vocoders at the MSC has distinct advantages over locating them at the base station. First, there is an extra level of compression that this allows on the link between the base station and the MSC, which can increase the call-carrying capacity of this link. Second, there is efficiency associated with the pooling of vocoders at the MSC that may be used across many base stations. Third, it allows the support of vocoder bypass in mobile-to-mobile calls, thus increasing the voice quality by removing vocoder tandeming. Fourth, there is a reduction in the processing load at the base station, which can lead to a less expensive base station implementation or can provide more processing power for other tasks performed by the base station. Fifth, there is more flexibility to implement new vocoders without requiring new software, more memory, or more processing power in every base station.

11.2 Call processing

Call processing includes the functions of call establishment and call handoff. The call establishment procedures vary depending upon whether the call is originated by a mobile or delivered to a mobile. The call handoff procedures vary depending upon whether it is an intrasystem or intersystem handoff, and whether the handoff is from an AVC or a DTC. Because Digital PCS allows for the definition of multiple vocoders, there is also a protocol for vocoder assignment that must be followed on a per-call basis.

11.2.1 Call establishment

The first step of call establishment is to determine from the mobile's HLR the calling services the mobile is allowed. This is called service qualification [13] and typically occurs when the mobile registers for service. The service qualification information for the mobile is stored in the serving system's VLR. The serving system determines through service qualification (1) the period for which the mobile is authorized for calls, (2) the types of calls the mobile is allowed to originate, (3) whether the mobile is allowed to have calls delivered to it, and (4) the calling features that the mobile is allowed. The call authorization period may be per call; for a number of calls; for a number of hours, days, or weeks; or indefinitely. The types of originated calls that are allowed may be none, local calls only, calls to a specified NPA-NXX only, calls to a specified NPA-NXX-XXX only, calls within the serving country, or all calls. Examples of calling features that a mobile may be allowed include call waiting, call forwarding, call transfer, three-way calling, calling number presentation, calling name presentation, and voice privacy.

An example of typical message flows for the call origination process is shown in Figure 11.5. For this example, it will be assumed that the mobile is in the DCCH Camping state when the call is originated. When the mobile powers up, finds a DCCH, and sends a registration message to the serving system, the serving MSC launches an TIA/EIA-41 registration notification (REGNOT) message to the mobile's HLR. The HLR returns the mobile's service qualification in the regnot message. This information is stored in the serving system's VLR record for the mobile.

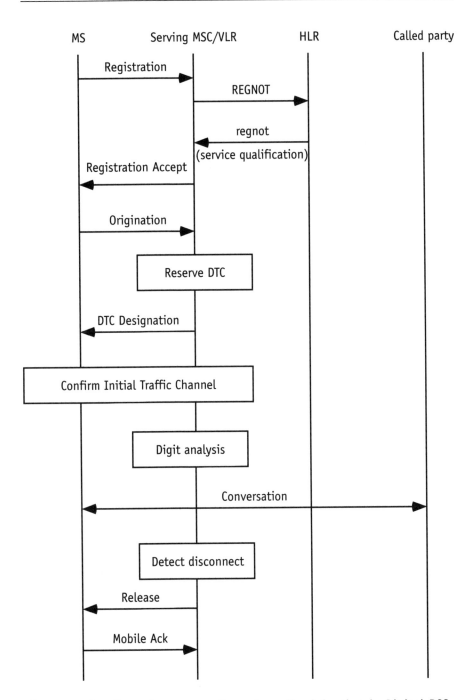

Figure 11.5 Example message flows for call origination in Digital PCS.

Not shown in the message flows is the signaling required for authentication of the mobile, which is covered in Chapter 16.

When the mobile user dials a directory number and presses the SEND key, the mobile launches an Origination message to the serving system on the reverse control channel. The MSC determines if there is an available DTC for the mobile, reserves one if available, and sends a Digital Traffic Channel Designation message to the mobile. Note that if a DTC is not available on the sector or cell from which the mobile originated the call, the MSC may send the mobile a Directed Retry message. This is a trigger condition for the mobile to reselect to a neighboring control channel and reoriginate the call, as described in Chapter 9. If a Digital Traffic Channel Designation message is received by the mobile, it tunes to the assigned channel and time slot and executes the Confirm Initial Traffic Channel task. During this task, the mobile acquires synchronization and time alignment with the forward DTC and begins transmitting at the assigned power level. The serving system confirms that the mobile is transponding the correct DVCC. It may also send the mobile Physical Layer Control messages to change its transmit power level and adjust its time alignment. At this point, the mobile enters the Conversation task on the DTC.

When the serving system has confirmed that the mobile is on the assigned channel, the MSC analyzes the dialed numbers. Based on the mobile's service qualification, the serving MSC determines if the mobile is allowed to originate a call to the dialed number. If it is, the serving MSC sets up the call to the destination and cuts through the voice path in the reverse direction to allow the calling party (the mobile user) to hear call process tones. When the trunk is answered, the serving MSC cuts through the voice path in the forward direction to allow the calling party to communicate with the called party, and a conversation may take place.

Figure 11.5 shows an example of the called party releasing the call. The serving MSC detects a trunk disconnect, releases the trunk, and sends a Release message to the mobile on the DTC. The serving MSC releases the DTC when the mobile acknowledges receipt of the Release message. The mobile returns to a control channel after sending the acknowledgement. The Release message may contain the DCCH Information information element, which the mobile uses to rapidly tune to a DCCH upon call release. Not shown in the figure is the message flow for a mobile release. In this case, the mobile user presses the END key and the mobile launches a release message on the DTC. The serving MSC releases the trunk to the called party and sends the mobile a Base Station Ack message, which may include DCCH information. The serving MSC then releases the DTC and the mobile tunes to a control channel.

An example of typical message flows for the call delivery (also called call termination) process is shown in Figure 11.6. It is assumed that the mobile is in the DCCH Camping state and registered for service on the

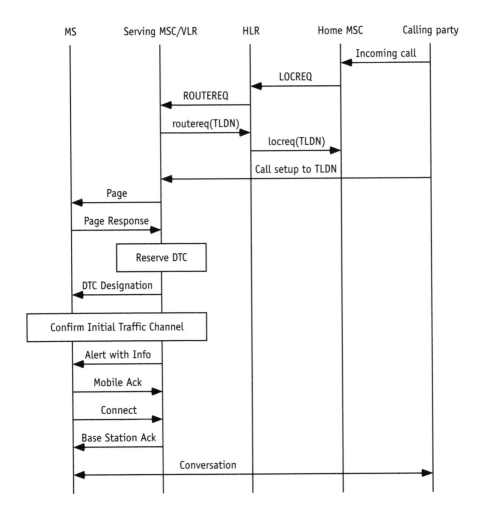

Figure 11.6 Example message flows for call termination in Digital PCS.

serving system when an incoming call arrives at the mobile's home MSC. The home MSC must query the HLR to determine the location of the mobile in order to route the call appropriately. To do this, the home MSC launches an TIA/EIA-41 Location Request (LOCREQ) message to the HLR, including the mobile's MSID in the message. The HLR has stored the destination point code for the serving system where the mobile has registered. It launches an TIA/EIA-41 route request (ROUTREQ) message to the serving MSC to determine how to route the call. The serving system allocates a temporary local directory number (TLDN) for the call and returns it to the HLR in a routreq message. The HLR uses the locreq message to relay the TLDN to the home MSC, and the home MSC routes the call to the TLDN.

Next, the serving system must page the mobile for the incoming call. It typically pages the mobile in the location area where the mobile last registered. If the mobile does not respond to the page within a few seconds, the serving system may page the mobile again in the location area, or may expand the scope of the paging to flood the entire network. The period between pages and the allowable number of pages per call attempt are typically parameters settable by the local service provider. A layer 2 hard page is typically used to page a mobile on the DCCH, as described in Chapter 5. The mobile is paged using one of its MSIDs, as described in Chapter 8. When the mobile receives a page addressed to it, it sends a Page Response message to the network and enters the Waiting for Order state.

When the serving system receives a Page Response message from the mobile, it stops paging the mobile and reserves a DTC in the appropriate cell or sector for channel assignment. If no channels are available in the cell or sector where the Page Response was received, the serving system may send the mobile a Directed Retry message, as described previously in this chapter. The serving system sends a Digital Traffic Channel Designation message to the mobile, providing the channel, time slot, DVCC, DMAC, and other parameters to use when accessing the DTC. When the mobile receives the message, it tunes to the assigned channel and time slot and executes the Confirm Initial Traffic Channel task. During this task the mobile acquires synchronization and time alignment with the forward DTC and begins transmitting at the assigned power level. The serving system confirms that the mobile is transponding the correct DVCC. It may also send the mobile Physical Layer Control messages to change its transmit power level and adjust its time alignment. At this point, the mobile enters the Conversation task on the DTC.

When the serving system has confirmed that the mobile is on the assigned channel, the serving system sends an Alert with Info message to the mobile on the DTC. This provides the mobile user with audible and visual alerting of the incoming call. The mobile sends the serving system a Mobile Ack message to confirm receipt of the Alert with Info message. If the mobile user decides to answer the call, the mobile sends a Connect message to the serving system, which acknowledges receipt of this message with a Base Station Ack and then proceeds to connect the call and cut through the voice path. Call release occurs exactly as described previously in this chapter for the call origination scenario.

11.2.2 Call handoff

One or more handoffs may be required while a mobile call is in progress. Mobile assisted handoff (MAHO) is used in Digital PCS to determine when a handoff is necessary. The handoff criteria are not standardized for Digital PCS and are specific to the network implementations. Many handoff parameters are under the control of the service provider to allow flexibility in network design. The signal procedures that facilitate handoffs are standardized in TIA/EIA-41 and TIA/EIA-136. An example of typical message flows for the handoff process is shown in Figure 11.7.

The network typically sends the mobile a Start Measurement Order message on the DTC, which is acknowledged by the mobile in a Measurement Order Ack message. The Start Measurement Order includes a set of up to 24 channels for the mobile to perform received signal strength (RSS) measurement on, then continuously report these measurements to the network in Channel Quality Measurement (CQM) messages. Included in the CQM messages are the RSS and bit error rate (BER) measurements on the current DTC. The network may send the mobile a Stop Measurement Order to stop the mobile measurements and CQM reports. This may be done in order to change the channels that the mobile is measuring. The MAHO channels sent to the mobile for measurement are typically the channels on which the DCCHs of neighboring cells and sectors are located. These channels are used because they are guaranteed

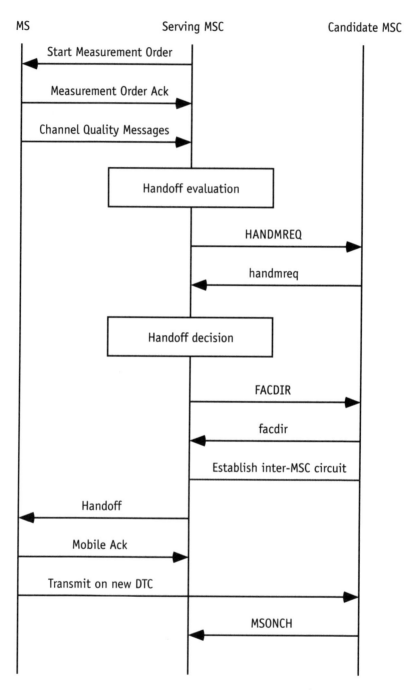

Figure 11.7 Example message flows for call handoff in Digital PCS.

to be continuously transmitting, so the mobile can measure RSS on them at any time.

Handoff algorithms in Digital PCS take advantage of the MAHO measurements, but they also use measurements made by receivers in the cellular network. Signaling to request different base stations to collect and report RSS measurements of a mobile's transmissions is only standardized for intersystem operation. For border areas where a handoff may be necessary to a cell or sector served by a different MSC, the TIA/EIA-41 handoff measurement request (HANDMREQ) message is used to request measurement data on a particular mobile. The candidate system reports the measurement results in the handmreq message.

The serving MSC uses the downlink and uplink measurements it has collected to attempt to place the mobile on the best server at all times. In fact, the handoff algorithm may be very similar to the reselection algorithm described in Chapter 9, which is used by the mobile to determine the best control channel to obtain service from at any given time. Similar handoff and reselection algorithms allow the service provider to more easily match handoff and reselection boundaries, as described in Chapter 17.

If the serving MSC determines that an intersystem handoff is desired, a TIA/EIA-41 facilities directive (FACDIR) message is sent to the target MSC requesting a call handoff. The target MSC notifies the serving MSC that it will accept the handoff in the facdir message, which includes handoff information necessary for the mobile to move to the new channel. An inter-MSC circuit is established between the serving and target MSCs, and the serving system sends the mobile a Handoff message on the current DTC. The mobile acknowledges receipt of the message, tunes to the new channel, and begins transmitting. Once the target system has confirmed the mobile is present on the assigned channel, it sends a mobile on channel (MSONCH) TIA/EIA-41 message to the serving system, which connects the other party to the mobile via the inter-MSC circuit.

11.2.3 Vocoder assignment

Digital PCS supports the definition of multiple vocoder that may be used in the mobile and the network. The mobile may be set to prefer one type of voice service over another, and may even allow the user to select analog voice as preferred. Signaling between the network and the mobile is used to negotiate the type of voice service used on a call-by-call basis.

When a mobile enters the DCCH Camping state, it reads the F-BCCH and E-BCCH subchannels. The Service Menu message may be broadcast by the network on one of these subchannels. This is an optional message, so the network is not obliged to broadcast it. In the absence of this message, the mobile is to assume the network is capable of analog speech and digital speech using the VSELP vocoder. If the service menu message is broadcast, a Voice Coder Map and a Menu Map are included in the message. The Voice Coder Map information element informs the mobile of the vocoders supported by the network, which would typically include IS-641. Among other things, the Menu Map information element informs the mobile if analog speech is supported.

Once the mobile reads the Service Menu message, it has all the information it needs to request the appropriate voice service in an Origination or Page Response message. If the mobile user originates a call prior to the mobile reading the Service Menu, the mobile must assume support for analog voice and digital voice with VSELP. (Because a mobile can originate a call from a DCCH after reading the F-BCCH and before reading the E-BCCH, it behooves the service provider to broadcast the service menu on the F-BCCH if IS-641 is the desired voice service.) When the mobile user originates a call or responds to a page, the service code and optionally the voice mode information element are included in the Origination or Page Response to request the desired voice service and vocoder. As long as the network has the desired resources available, it can honor the request in the Digital Traffic Channel Assignment or Analog Voice Channel Assignment message.

The target system in an intersystem handoff must know the capabilities of the mobile in order to assign the appropriate voice resources upon handoff. The serving system provides this information to the target system in a TIA/EIA-41 Facilities Directive 2 (FACDIR2) message [14]. The FACDIR2 message is similar to the FACDIR message, but contains additional information that includes the voice services capabilities of the mobile. The Time-Slot Indicator information element in the Handoff message is used to indicate the service code (analog or digital) on the new channel. The Dedicated DTC Handoff message may be used to change the voice mode (the vocoder) if the Voice Mode information element is included. If this information element is not included in the dedicated DTC Handoff message, or if the Handoff message is used and the Time-Slot Indicator indicates digital voice service, the mobile assumes the vocoder type used on the current DTC will continue to be used on the new channel.

References

[1] Chitrapu, P., "Modern Speech Coding Techniques and Standards," *Multimedia System Design*, Vol. 2, No. 2, Feb. 1998, p. 33.

[2] Ibid., pp. 22–24.

[3] Flanagan, J. L., et al., "Speech Coding," *IEEE Transactions on Communications*, Vol. COM-27, No. 4, Apr. 1979, p. 725.

[4] Winter, E., "Speech Coding," *Communication Systems Design*, Vol. 2, No. 5, May 1996, p. 20.

[5] Richey, R., and A. Lovrich, "Designing Low-Cost Speech Analysis and Synthesis," *Multimedia System Design*, Vol. 2, No. 4, Apr. 1998, p. 44.

[6] Cox, R. V., "Three New Speech Coders from the ITU Cover a Range of Applications," *IEEE Communications Magazine*, Vol. 35, No. 9, Sept. 1997, Figure 1, p. 42.

[7] Schroeder, M. R., and B. S. Atal, "Code-Excited Linear Prediction (CELP): High-Quality Speech at Very Low Bit Rates," *Proc. of ICASSP'85*, 1985, pp. 937–940.

[8] Laflamme, C., et al., "16 kbps Wideband Speech Coding Technique Based on Algebraic CELP," *Proc. of ICASSP'91*, 1991, pp. 13–16.

[9] Telecommunications Industry Association, *TIA/EIA/IS-641, TDMA Cellular/PCS—Radio Interface—Enhanced Full-Rate Speech Codec*, May 1996, pp. 5–7.

[10] Benyassine, A., et al., "ITU-T Recommendation G.729 Annex B: A Silence Compression Scheme for Use with G.729 Optimized for V.70 Digital Simultaneous Voice and Data Applications," *IEEE Communications Magazine*, Vol. 35, No. 9, Sept. 1997, pp. 64–73.

[11] Ibid., p. 64.

[12] Harte, L., *Dual Mode Cellular*, Bridgeville, PA: P. T. Steiner Publishing Co., 1992, pp. 6-20 to 6-22.

[13] Gallagher, M. D., and R. A. Snyder, *Mobile Telecommunications Networking with IS-41*, New York: McGraw-Hill, 1997, pp. 166–168.

[14] Telecommunications Industry Association, *TIA/EIA-IS-730, IS-136 (DCCH) Support in IS-41*, July 1997, pp. 15–17.

12
Teleservices

12.1 Introduction

Teleservices were defined and briefly explained in Chapter 2. They are applications that use the air interface and network interface as bearers for transport between teleservice servers and mobiles to provide end-to-end services to users. Teleservices provide a flexible mechanism for service providers to deliver teleservice data units composed of user data or programming to mobiles that are operating on either the DCCH or DTC while in the home system or roaming. Except for some aspects of signaling, teleservice delivery is invisible to most of the cellular network because the teleservice server and mobile are the primary entities that produce and consume teleservices.

Teleservices reside above the network layer in the Open Systems Interconnection (OSI) reference model, as shown in Figure 12.1. Teleservices may contain some portions or all of the transport, session, presentation, and application layers. They may also be carried on top of a

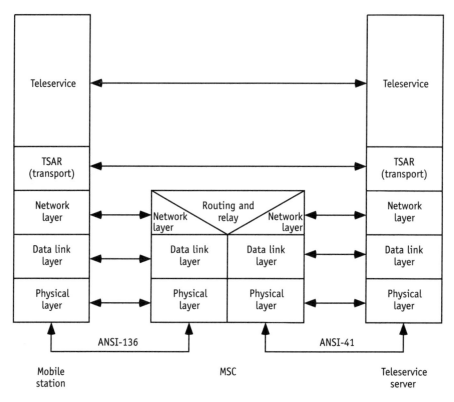

Figure 12.1 Where teleservices fit in the protocol stack.

separate transport layer, as depicted in the figure. The transport layer shown in the figure is the teleservice segmentation and reassembly (TSAR) function defined for TIA/EIA-136-A. Another portion of the transport layer that could be used for the delivery of teleservices is the broadcast teleservice transport function, also defined for TIA/EIA-136-A. This function provides for efficient broadcast and reception of teleservices across the air interface. Note that teleservices may either be point-to-point or broadcast in nature.

Teleservices used with Digital PCS may be defined or carrier-specific. A defined teleservice is one that is entirely described in one or more of the Digital PCS standards. A carrier-specific teleservice is one that may or may not be entirely described in a Digital PCS standard. Carrier-specific teleservices are typically used when rapid implementation is desired, or to provide differentiated services to users. Examples of defined teleservices for Digital PCS include cellular messaging teleservice, over-the-air activation teleservice, over-the-air programming teleservice, and general UDP transport service.

12.2 Teleservice delivery

Teleservices are carried across the TIA/EIA-41 network interface from the teleservice server to the serving MSC within the Short Message Delivery—Point-to-Point (SMDPP) message, and across the TIA/EIA-136 air interface from the base station to the mobile within the R-DATA message, as shown in Figure 12.2 [1,2]. The MSC translates the teleservice information between the SMDPP and R-DATA message formats. It also provides network layer acknowledgement to the originator (either the teleservice server or the mobile) upon successful delivery of a teleservice.

The teleservice server provides formatting, storage, and forwarding of teleservices. Teleservice data unit information to be included in teleservices destined for one or more mobiles is provided to the teleservice server by a short message entity (SME). Examples of SMEs include computers, interactive voice response (IVR) systems, other teleservice servers, and intelligent peripherals. Besides the serving MSC, the teleservice

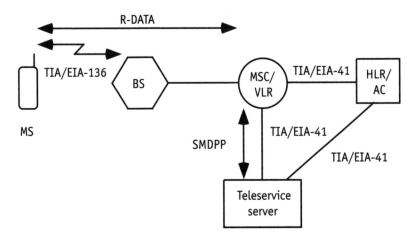

Figure 12.2 Network model for teleservice delivery.

server must communicate with the mobile's home location register (HLR) to obtain routing information for teleservices. These network elements and their functions are described in greater detail in Chapter 7. The exact message flows for the delivery of teleservices are unique to each teleservice.

The contents of the SMDPP and R-DATA messages are summarized in Tables 12.1 and 12.2, respectively. It can be seen from a review of the tables that there is a one-to-one mapping between many of the SMDPP parameters and the R-DATA information elements and fields. The SMS_BearerData maps to the Higher Layer Protocol Data Unit. The SMS_TeleserviceIdentifier maps to the Higher Layer Protocol Identifier. The SMS_OriginalDestinationAddress and SMS_OriginalDestination-Subaddress map to the User Destination Address and User Destination Subaddress, respectively. The SMS_OriginalOriginatingAddress and SMS_OriginalOriginatingSubaddress map to the User Originating Address and User Originating Subaddress, respectively. The serving MSC performs these translations. Note, however, that the serving MSC does not decode or interpret the SMS_BearerData or Higher Layer Protocol Data Unit.

Positive or negative acknowledgement of the delivery of a teleservice is automatically provided to the SME originating the teleservice. Across the network interface, the acknowledgement is provided within the TIA/EIA-41 SMDPP Return Result message (called smdpp, for short). Across the air interface, the acknowledgement is provided within either the R-DATA ACCEPT or R-DATA REJECT message. If the delivery of the teleservice was unsuccessful, then a negative acknowledgement and the cause of the rejected teleservice delivery are included within the smdpp and R-DATA REJECT messages. The SMS_CauseCode parameter within the smdpp message is used to indicate a reason for not delivering an SMS message. TIA/EIA-41 groups these reject causes into the categories of network problems, terminal problems, radio interface problems, and general problems, with specific reject causes under each category. TIA/EIA-136 has an information element called R-Cause that is included in the R-DATA REJECT message and corresponds to the SMS_CauseCode parameter. Examples of R-Cause values include "R-DATA message too long," "Memory capacity exceeded,"

Table 12.1
Summary of SMDPP Message Contents

SMDPP Parameter	Meaning
SMS_BearerData	The teleservice data unit (mandatory)
SMS_TeleserviceIdentifier	An indicator of the teleservice for which the teleservice data unit applies (mandatory)
Electronic Serial Number	The electronic serial number (ESN) of the originating mobile (optional)
Mobile Identification Number	The mobile identification number (MIN) of the originating mobile (optional)
SMS_ChargeIndicator	An indicator to specify various charging options for a teleservice message (optional)
SMS_DestinationAddress	The temporary routing address for the delivery of the teleservice (optional)
SMS_MessageCount	The number of teleservice messages pending delivery (optional)
SMS_NotificationIndicator	An indicator used to control the sending of subsequent SMS notification messages (optional)
SMS_OriginalDestinationAddress	The address of the original message destination (optional)
SMS_OriginalDestinationSubaddress	The subaddress of the original message destination (optional)
SMS_OriginalOriginatingAddress	The address of the original message sender (optional)
SMS_OriginalOriginatingSubaddress	The subaddress of the original message sender (optional)
SMS_OriginatingAddress	The current routing address of the originating mobile (optional)

"Destination out of service," "Invalid message," "Invalid information element content," and "Protocol error." The serving MSC maps the R-Cause values to the most appropriate SMS_CauseCode. Depending upon the

Table 12.2
Summary of R-DATA Message Contents

R-DATA Information Element	Meaning
Protocol Discriminator	The identification of the standard that defines the message (mandatory)
Message Type	An indicator that defines how to interpret the remaining information elements by indicating that this is an R-DATA message (mandatory)
R-Transaction Identifier	A number used to associate an R-DATA ACCEPT or R-DATA REJECT message with the R-DATA message (mandatory)
R-DATA Unit	The teleservice identity and teleservice data unit, composed of three fields, the Length Indicator, Higher Layer Protocol Identifier, and Higher Layer Protocol Data Unit (mandatory)
Length Indicator	An indicator of the remaining length of the R-DATA unit (mandatory)
Higher Layer Protocol Identifier	An indicator of the teleservice for which the teleservice data unit applies (mandatory)
Higher Layer Protocol Data Unit	The teleservice data unit (mandatory)
Teleservice Server Address	The address (e.g., point code and subsystem number) of the teleservice server that is either delivering the teleservice or receiving the teleservice (optional)
User Destination Address	The address of the original message destination (optional)
User Destination Subaddress	The subaddress of the original message destination (optional)
User Originating Address	The address of the original message sender (optional)
User Originating Subaddress	The subaddress of the original message sender (optional)
User Originating Address Presentation Indicator	An identifier of the presentation restrictions and screening related to the user originating address (optional)

reject cause, the originator of the teleservice may or may not retry sending the teleservice.

12.3 SMS_TeleserviceIdentifier and Higher Layer Protocol Identifier

The TIA/EIA-41 SMS_TeleserviceIdentifier parameter and the TIA/EIA-136 Higher Layer Protocol Identifier (HLPI) field both indicate the teleservice for which the teleservice data unit data applies. As described previously, the serving MSC translates the SMS_Teleservice-Identifier to the HLPI for teleservice delivery to a mobile, and from the HLPI to the SMS_TeleserviceIdentifier for teleservice delivery to the teleservice server. The SMS_TeleserviceIdentifier is 16 bits in length, while the HLPI is only 8 bits in length. The SMS_TeleserviceIdentifier must be longer than the HLPI because TIA/EIA-41 must support air interfaces other than TIA/EIA-136, which could have their own teleservices that are different than those designed for Digital PCS. The SMS_TeleserviceIdentifier is translated into the HLPI simply by ignoring the eight most significant bits, which are all set to 1 for TIA/EIA-136. The mobile and the teleservice server use the HLPI to determine how to interpret the teleservice data unit.

The HLPI is divided into three subfields, as shown in Figure 12.3. The most significant bit is the Teleservice Type Indicator, which indicates if the teleservice is a defined one or is carrier-specific. The next most

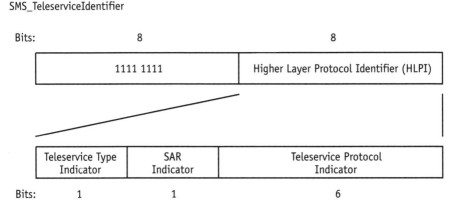

Figure 12.3 The formatting of an HLPI.

significant bit is the Segmentation and Reassembly Indicator, which identifies if teleservice segmentation and assembly has been applied to the teleservice data unit. The third subfield is the Teleservice Protocol Identifier, which is composed of the least significant six bits of the HLPI and defines the actual teleservice included in the teleservice data unit. The teleservice data unit is called the Higher Layer Protocol Data Unit in TIA/EIA-136.

When a mobile receives an R-DATA message, it decodes the message up to the point where the HLPI can be identified. It determines if the HLPI is one that it supports before attempting to decode the Higher Layer Protocol Data Unit. This is necessary because the rules for decoding and interpreting the Higher Layer Protocol Data Unit depend on the specific teleservice. The mobile checks the Teleservice Protocol Identifier and the Teleservice Type to determine if it supports the received teleservice. Teleservice Protocol Identifiers defined for Digital PCS are shown in Table 12.3. These teleservices may be standardized or carrier-specific, based on the value of the Teleservice Type Indicator. The mobile sends an R-DATA REJECT message to the network with the appropriate cause code (protocol error) if it does not support the received teleservice. Otherwise, it begins decoding the Higher Layer Protocol Data Unit according to the rules governing the teleservice.

Table 12.3
Defined Teleservice Protocol Identifiers

Teleservice Protocol ID	Teleservice
00 0000	Network-specific (not currently used in Digital PCS)
00 0001	Cellular messaging teleservice
00 0010	Cellular paging teleservice (not currently used in Digital PCS)
00 0011	Over-the-air activation teleservice
00 0100	Over-the-air programming teleservice
00 0101	General UDP transport service

12.4 Teleservice segmentation and reassembly

The maximum size of the Higher Layer Protocol Data Unit in an R-DATA message is limited to 251 octets across the TIA/EIA-136 air interface. Similarly, the maximum size of the SMS_BearerData in an SMDPP message is limited to 212 octets if SS7 is used as the TIA/EIA-41 signaling protocol. Optional information elements and parameters that are normally sent with a teleservice limit the size of the usable data even further. It is desirable to send messages that are longer than a single R-DATA or SMDPP message can accomodate. TSAR makes this possible. TSAR allows the originator of a teleservice to segment a long message into smaller chunks for delivery across the network interface and air interface to a destination where the segments are reassembled into the original long message.

TSAR works by defining Begin, Continue, and End messages that contain the segments of the teleservice. If a teleservice message can fit in one SMDPP and R-DATA message, TSAR is not applied. If two SMDPP and R-DATA messages are needed, the teleservice message is broken into two segments and sent as a Begin message and an end message. If three segments are needed, one Begin, one Continue, and one End message are sent. If more than three segments are needed, multiple Continue messages are sent. Each TSAR message contains a sequence number and a transaction ID to allow a mobile to begin receiving a segmented teleservice message in midstream. This capability is particularly useful for broadcast teleservices, in which a mobile may begin reading a subchannel after the Begin message has been transmitted.

12.5 Broadcast teleservice transport

Broadcast teleservice transport provides an efficient way of delivering broadcast teleservices to groups of mobiles. Among other things, it provides the capability to define zones within a cellular network where different teleservices may be broadcast, segment broadcast teleservices into categories and service groups, and provide overhead information to mobiles that allows them to more efficiently read broadcast teleservices of interest without significantly impacting standby time. Many of the

Digital PCS-defined teleservices may be delivered in either a broadcast or point-to-point manner. Broadcast teleservice delivery can be a more efficient way for a cellular network to deliver teleservices to mobiles if the same information is desired by many mobiles.

Broadcast teleservice transport is composed of three types of data: broadcast teleservice transport information (BTTI), broadcast overhead information, and the broadcast teleservices. Some of the BTTI is consumed by the cellular network and some becomes a part of the broadcast overhead information. The BTTI is information that defines the broadcast attributes of teleservices. The broadcast overhead information is used to inform the mobile of the specific categories of teleservices that are broadcast and the subchannels that carry these categories. The broadcast overhead information can also be used to inform mobiles that a change has occurred to one or more of the broadcast teleservices. The broadcast teleservice is the actual teleservice that is encapsulated by the SMDPP and R-DATA messages, as described above.

An example of the delivery of broadcast teleservices to a group of mobiles using broadcast teleservice transport is shown in Figure 12.4. A broadcast teleservice is originated by a teleservice service, which delivers the broadcast teleservice to the serving MSC in one or more SMDPP messages. The teleservice server includes the BTTI in the SMDPP messages, and the serving MSC uses this information to identify the cluster of base stations that should broadcast the teleservice and the subchannel upon which the teleservice should be broadcast. The serving MSC also repackages part of the BTTI and the teleservice information contained in the SMDPP message into broadcast overhead and one or more R-DATA messages for transmission across the Digital PCS air interface. The DCCH is used to transmit the broadcast overhead and broadcast teleservice information to mobiles. Forward DCCH slots may be defined as an SMS broadcast channel (S-BCCH) for this purpose. Mobiles capable of receiving broadcast teleservices read the broadcast overhead information on the S-BCCH and identify subchannels of the S-BCCH that contain categories of interest to them. They may then read these subchannels to obtain the broadcast teleservice. Unlike point-to-point delivery of teleservices, there is no acknowledgement of successful delivery of broadcast teleservices to an SME.

Teleservices   215

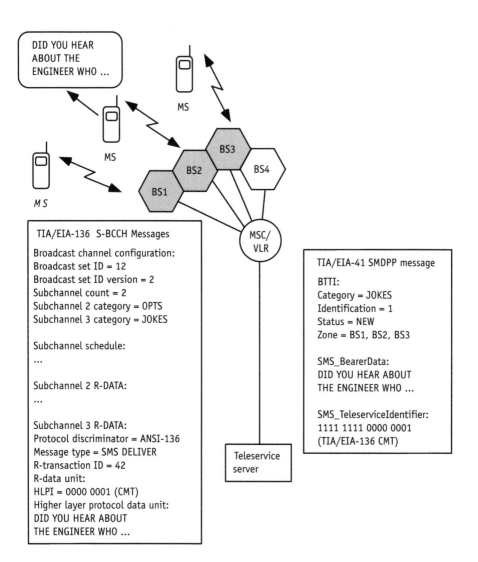

Figure 12.4 How broadcast teleservices are delivered to mobiles using broadcast teleservice transport.

Table 12.4 lists the BTTI parameters that may be delivered to the serving MSC in the SMDPP message from the teleservice server. Some of this information does not apply to Digital PCS, but to other air interface technologies. Of particular interest for Digital PCS are the Category, Status, Group, and Zone parameters. The Category parameter is used to select the subchannel of the S-BCCH upon which the teleservice is broadcast. The Category may be as simple as the teleservice identification (for example, cellular messaging teleservice and over-the-air programming teleservice), or may be more specific (for example, news, weather, and sports). The Status parameter is used to add or remove teleservice messages from the broadcast subchannels. The Group parameter may define the destination service group for the broadcast teleservice. As described previously, the Zone parameter defines the cluster of base stations from which a particular teleservice is to be broadcast. While the Encryption parameter may not have meaning in Digital PCS, broadcast teleservices may be encrypted end-to-end. That is, encryption is applied to the SMS_BearerData at the teleservice server and the Higher Layer Protocol Data Unit contents are decrypted by mobiles that have access to the decryption key.

Table 12.4
BTTI Summary

BTTI Parameter	Meaning
Category	Indication of the specific subject matter of the teleservice payload
Identification	Unique identification of a message
Encryption	Indication of whether encryption should be used to protect the contents of a broadcast message and to prevent access by unauthorized subscribers
Status	Indication of whether the message is new, a replacement, or a deletion of an existing message with the same identification
Priority	Indication of the level of priority of the broadcast message
Character Set	Identification of the character set used in the message
Group	Identification of the target mobile audience
Zone	Indication of the geographical area over which the message should be broadcast
Periodicity	Indication of the start time, duration, and repetition rate of the broadcast

The broadcast overhead information includes the configuration of the broadcast subchannels, the schedule for broadcast subchannel transmission, change notifications, and teleservice messages sent to describe message categories that are not predefined. This broadcast overhead information assists the mobile in identifying the nature of the broadcast messages on the S-BCCH. The S-BCCH may contain two subchannels devoted to broadcast overhead information and up to 30 subchannels that carry teleservice messages. The configuration of the broadcast channel, sent in the Broadcast Channel Configuration message, identifies the broadcast set for the S-BCCH, the version of the broadcast set, and the category of information contained in each subchannel that carries teleservice messages. The schedule for broadcast subchannel transmission, sent in the Subchannel Schedule message, identifies when each subchannel will be broadcast. The Change Notification message identifies if there is a change to one or more of the broadcast subchannels, and the messages that have been changed.

The broadcast overhead information helps mobiles conserve battery life. The mobile can rapidly read this information to determine (1) if there is a category of interest that is broadcast on the S-BCCH, (2) exactly when in time a category of interest will be broadcast, (3) if there has been a change to any of the categories of interest that have already been read, and (4) if this S-BCCH is broadcasting the same teleservice information as another S-BCCH on a different DCCH that has already been read. This keeps the mobile from reading extraneous information or rereading information that has not changed.

When a mobile determines from reading the broadcast overhead information that the S-BCCH is broadcasting a category of interest to it, the mobile proceeds to read the appropriate subchannel and process the teleservice. If the category is text-based, such as the cellular messaging teleservice, the mobile displays the broadcast information to the user. If the category is programming, such as the over-the-air programming teleservice, the mobile updates its programming accordingly.

12.6 Defined teleservices

This section describes the defined, or standardized, teleservices for Digital PCS.

12.6.1 Cellular messaging teleservice

Cellular messaging teleservice (CMT) is the teleservice used to provide short message service (SMS) to mobile users. With CMT, alphanumeric messages can be sent to and from mobiles, thus providing a functionality equivalent to an alphanumeric pager. In the past, a person who had to be reachable all the time would typically wear a pager and carry a mobile. When he or she received a message on the pager, he or she would use the mobile to respond to the page. The use of CMT to provide SMS allows a person to eliminate the need to wear a pager because the mobile acts as the pager.

CMT may be used to deliver mobile-terminated SMS (MT-SMS) and mobile-originated SMS (MO-SMS). MT-SMS is one-way only, allowing for alphanumeric message delivery from the network to the mobile. The mobile automatically sends an acknowledgement upon receipt of an MT-SMS message to inform the network that the message has been received. MT-SMS and MO-SMS together form a two-way data service that can be used to allow mobile users to respond to messages delivered to them or send messages to others. Short messages originated by the mobile may be entered from the mobile keypad, from an attached computer, or from a set of canned messages stored in the mobile. TSAR may be applied to a CMT message in order to send a message longer than what is supported in a single TIA/EIA-41 and TIA/EIA-136 message. End-to-end encryption and separate message encryption across the air interface may be applied to a CMT message to increase its security. A CMT message may be carried over a DCCH or a DTC to a mobile. CMT messages may also be carried across the S-BCCH as broadcast teleservice messages.

There are four key CMT messages. The SMS DELIVER message is sent from the teleservice server to the mobile, and the SMS SUBMIT message is sent from the mobile to the teleservice server. The mandatory and optional information elements contained in the SMS DELIVER and SMS SUBMIT messages are summarized in Table 12.5. Two additional CMT messages may be used for acknowledgements. The SMS DELIVERY ACK and SMS MANUAL ACK messages may be sent in response to a SMS DELIVER or SMS SUBMIT message when delivery acknowledgement and manual acknowledgement, respectively, are requested in these messages. The SMS DELIVERY ACK is sent when the

Table 12.5
Information Elements for SMS DELIVER and SMS SUBMIT CMT Messages

Information Element	SMS DELIVER	SMS SUBMIT	Meaning
Message Type Indicator	Mandatory	Mandatory	The CMT message type (e.g., SMS DELIVER, SUBMIT, SMS DELIVERY ACK, or SMS MANUAL ACK).
Message Reference	Mandatory	Mandatory	An integer representation of a reference number for a short message, used for tracking and acknowledgement purposes.
Privacy Indicator	Mandatory	Mandatory	An indicator of the level of privacy (not restricted, restricted, confidential, or secret) of the received message. The receiving terminal may process a message differently depending upon the level of privacy.
Urgency Indicator	Mandatory	Mandatory	Identifies the level of urgency associated with the received message (bulk, normal, urgent, very urgent). The receiving terminal may process a message differently depending upon the level of urgency.
Delivery Acknowledgement Request	Mandatory	Mandatory	An indicator that delivery acknowledgement is requested or prohibited.
Manual Acknowledgement Request	Mandatory	Mandatory	An indicator that manual acknowledgement is requested or prohibited.
Message Updating	Mandatory	Mandatory	An indicator that identifies to the receiving terminal whether or not the current message overwrites a previously received message.
Validity	Mandatory	Not sent	Identifies the period of time for which the received message is valid (indefinitely, until power down, while registered with the current system, or display only).

Table 12.5 (continued)

Information Element	SMS DELIVER	SMS SUBMIT	Meaning
Display Time	Mandatory	Not sent	An indicator to the receiving mobile of the time that the message should be displayed (Temporary, Default, or Invoke).
User Data Unit	Mandatory	Mandatory	The actual user text message.
Validity Period	Not sent	Optional	An indicator to the teleservice server of the relative or absolute time after which the message could be deleted if not delivered to the intended recipient.
Deferred Delivery Time	Not sent	Optional	The time the teleservice server should wait after receiving an SMS SUBMIT message before delivering the message to the recipient, specified as either a relative time from when the message is received by the teleservice server or an absolute time.
Message Center Time Stamp	Optional	Not sent	An indication of the time when the teleservice server received the short message.
SMS Signal	Optional	Not sent	The pitch, cadence, and duration that should be used to alert the user upon receipt of the message.
Callback Number	Optional	Optional	The callback number associated with the message. The callback number may be a directory number, Internet address, or another form of addressing. Multiple callback numbers may be provided.
Callback Number Presentation Indicator	Optional	Optional	The presentation and screening indications specific to the associated instance of the callback number information element. Each callback number has its own presentation and screening indications.
Callback Number Alpha Tag	Optional	Optional	The alphanumeric tag specific to the associated instance of the callback number information element. Each callback number has its own alphanumeric tag, which may be up to 64 characters in length.

CMT message is opened by the recipient. This is much like the capability available with some electronic mail systems for the originator of an e-mail message to receive a notification when a sent message is read. The SMS MANUAL ACK message is sent when the recipient provides a predetermined response to the CMT message. An example would be for the recipient to respond to a CMT message, "Can you meet me at 10 AM in my office? 1 = Yes, 2 = No," with a response code 1, indicating "Yes."

Figure 12.5 shows message flows for both MT-SMS and MO-SMS. The example has been formulated to describe most of the signaling that could occur in the process of short message delivery. A mobile user enters a short text message into mobile 1 (MS1) and addresses it to the

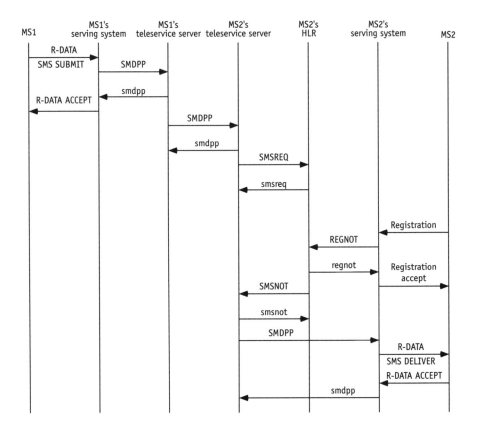

Figure 12.5 Message flows for MT-SMS and MO-SMS delivery.

mobile station identity (MSID) of mobile 2 (MS2). MS1 encapsulates this short text message into the user data unit of an SMS SUBMIT message. The SMS SUBMIT message is delivered to the cellular network in an R-DATA message on either a DCCH or DTC. The serving system translates the R-DATA message into an SMDPP message and delivers it to MS1's teleservice server. MS1's teleservice server acknowledges receipt of the short message to the serving system in an smdpp message, and the serving system sends the mobile an R-DATA ACCEPT message to confirm the short message has made it to the mobile's teleservice server.

At this point, what appears to be a miracle occurs and MS1's teleservice server routes the short message to MS2's teleservice server in an SMDPP message. Actually, it is not a miracle, but can be accomplished in either of two ways. First, MS1's teleservice server could contain a routing table that identifies the destination point code of the teleservice server corresponding to MS2's MSID. Second, the SS7 global title translation (GTT) capability could be used, in which MS1's teleservice server creates a global title address containing MS2's MSID and requests an MSID-to-teleservice server global title translation. An SS7 signaling transfer point translates the global title into the destination point code of MS2's teleservice server, and the SMDPP message is routed appropriately. MS2's teleservice server responds to the SMDPP message with an smdpp message to MS1's teleservice server.

If MS2's teleservice server does not have a current routing address for MS2, it sends an SMS Request (SMSREQ) message to MS2's HLR requesting the routing address and a notification when MS2 becomes available. Since MS2 is unavailable, the HLR informs the teleservice server in the smdpp message that delivery of the short message must be postponed. The HLR also sets an SMS Delivery Pending flag for MS2. This flag indicates that MS2's teleservice server should be notified when MS2 becomes available. At some later time, MS2 sends a Registration message to a serving system (a home or roaming system). The serving system sends a REGNOT message to MS2's HLR, including a routing address for short messages. When the HLR responds with a regnot, the serving system send MS2 a Registration Accept message.

Since the SMS Delivery Pending flag is set for MS2, the HLR launches an SMS Notification (SMSNOT) message to MS2's teleservice server including the routing address for short messages destined for MS2. The teleservice server sends the HLR an smsnot message, then sends an SMDPP message to the serving system with the short message for MS2 included in an SMS DELIVER CMT message. The serving system translates the SMDPP message into an R-DATA message and sends it to MS2. MS2 acknowledges receipt of the short message by sending an R-DATA ACCEPT message to the serving network. The serving network sends an smdpp message to MS2's teleservice server acknowledging the short message was delivered to the mobile, and the delivery process is completed.

12.6.2 Over-the-air activation teleservice

Over-the-air activation teleservice (OATS) is the teleservice used to provide over-the-air activation (OAA). OAA is a method for provisioning a mobile user with the home service provider's network and programming the mobile's NAM with the mobile station ID (MSID), the home service provider identity (SID and SOC), the A-Key, and other information. OAA can be accomplished without requiring the mobile user and the service provider to be in the same location. This opens up new opportunities for service providers to distribute mobiles and provision them for service.

OATS requires two-way R-DATA capability. TSAR may be applied to an OATS message in order to send a message longer than what is supported by individual TIA/EIA-41 and TIA/EIA-136 messages. The most important OATS messages are encryption across the air interface. An OATS message may be carried over a DCCH or a DTC to a mobile. Since OAA is performed on an individual mobile, OATS messages are not carried across the S-BCCH as broadcast teleservice messages.

OATS messages and the OAA process are best explained through an example of the message flow required to activate an unprogrammed mobile. The network architecture for the delivery of teleservices shown in Figure 12.2 is supplemented with additional elements and functions to describe the OAA process. The supplemented network architecture and example message flow are shown in Figures 12.6 and 12.7, respectively. Added to the network architecture is the customer service center (CSC), which launches the activation process and enters subscriber information into the HLR and billing system. Also added to the network architecture

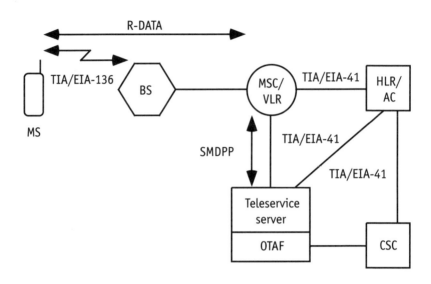

Figure 12.6   Network architecture for OAA.

is an over-the-air activation function (OTAF), a functional part of the teleservice server that handles OATS messages and serves as a temporary HLR for an unactivated mobile.

An unactivated mobile supporting OATS uses an activation MIN as its MSID when accessing the system. The activation MIN has NPA = 000, and NXX-XXXX = 7 least significant decimal digits of the ESN (see Chapter 8 for a description of these MIN fields). When the unactivated mobile powers up, finds a DCCH, and sends a Registration message, it uses this activation MIN to identify itself to the serving system. The serving MSC contains a routing table that points to the OTAF as the HLR for all mobiles with NPA = 000. The serving MSC formulates a Registration Notification (REGNOT) message and sends it to the OTAF. The OTAF sends a regnot to the serving MSC with the Origination Indicator and Allowed Call Types parameters set to restrict the unprogrammed mobile from originating any call except to the activation number for the service provider.

When the user of the unactivated mobile originates a call, the mobile sends an Origination message to the serving system, which sends a DTC Designation message to the mobile and routes the call to the CSC. The

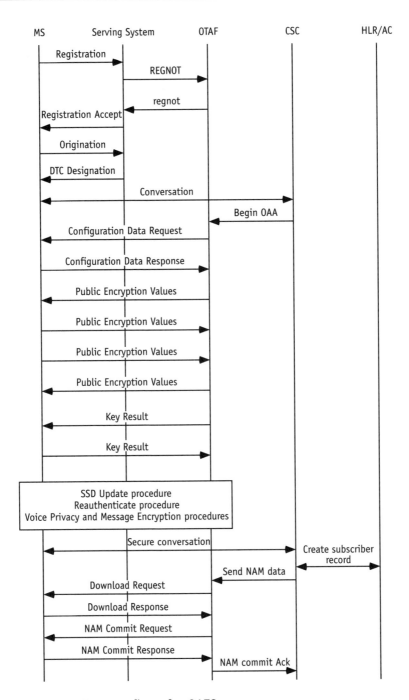

Figure 12.7 Message flows for OATS.

user and a customer service representative for the service provider converse, and it is determined that the user desires to activate the mobile on the service provider's network. The CSC launches a message to the OTAF to begin the programming process. For the remainder of this description, whenever an OATS message is sent, only the OATS message will be referred to; the SMDPP and R-DATA messages encapsulate the OATS messages.

The OTAF sends a Configuration Data Request message to the mobile. This message requests the mobile to send the OTAF a copy of its current NAM data. The mobile responds with a Configuration Data Response message providing its NAM data. This transaction is not required for the initial activation process, but is useful for reprogramming a mobile's NAM. Not shown in the message flow is a CSC challenge/response procedure that the mobile may use to verify that the CSC has programming privileges. This CSC challenge/response procedure is based on knowledge of a secret key stored in the mobile and the CSC/OTAF authorized to program the mobile.

The next step in the OAA process is for the HLR/AC and the mobile to generate an A-Key, using the OTAF to relay OATS messages that accomplish this. Depending upon the implementation, the OTAF may perform all of the A-Key generation and temporarily store the A-Key until the mobile's permanent HLR/AC is identified, then transfer the A-Key to the mobile's HLR/AC in a secure manner. This is depicted in the figure. The mobile's A-Key is generated using public key encryption; the A-Key is *not* transmitted over the air. Public key encryption allows the network to generate a secret key and calculate a network public key that is shared with the mobile in a Public Encryption Values message. The mobile also generates a secret key, and it uses this secret key and the network public key to generate a mobile public key that is shared with the network in a Public Encryption Values message. From the public keys that are shared and the secret keys that are not shared, both the OTAF and the mobile can calculate the same A-Key. The Key Result message is sent by both the OTAF and the mobile when the A-Key has been generated. A fraudster attempting to gain access to the A-Key for the mobile as it is generated will only have knowledge of the public keys that have been exchanged, which is not enough information for him to deduce the A-Key.

Once the A-Key has been generated by both the OTAF and the mobile, the OTAF initiates the Shared Secret Data (SSD) Update procedure. The signaling to accomplish the SSD update is described in Chapter 16, and is not duplicated here. This is necessary so the mobile and the OTAF are in possession of the same SSD for the mobile. The OTAF next initiates a Reauthentication procedure, which is very similar to the unique challenge authentication process described in Chapter 16. The purpose of this Reauthentication procedure is to ensure the mobile has stored the appropriate authentication-related variables that are used to generate the voice privacy mask and signaling message encryption key. When the reauthentication procedure is successfully completed, voice privacy and signaling message encryption are activated. At this point, the conversation between the mobile user and the customer service representative is secure and billing information may be obtained. The remainder of the OATS messages are also encrypted to protect their contents from fraudsters.

Once the customer service representative has acquired enough information to build a subscriber record and activate the mobile, the CSC creates the subscriber record in the HLR and sends the appropriate NAM data to the OTAF for relay to the mobile. The OTAF formulates a Download Request message and sends it to the mobile. The mobile stores this information in temporary NAM and sends a Download Response message to the OTAF to verify temporary storage. The OAA process is completed when the OTAF sends a NAM Commit Request message to the mobile, the mobile permanently stores the NAM data, and the mobile sends a NAM Commit Response message to the OTAF verifying that the NAM data has been made permanent.

12.6.3 Over-the-air programming teleservice

Over-the-air programming teleservice (OPTS) is the teleservice used to provide over-the-air programming (OAP). OAP is a method for programming non-NAM data into a mobile (OAA is used to program NAM data). The first application for OAP is the programming of the intelligent roaming database (IRDB) into mobiles (see Chapter 15). However, OAP could be enhanced in the future to program additional information into mobiles, such as new call processing or teleservice logic. Such programming could be used to correct bugs in the operation of mobiles, or to provide mobiles with the ability to support new features and services.

Currently, OPTS is defined as a mobile-terminated teleservice between an OTAF and a mobile. The OTAF formulates an OPTS message, encapsulates it in an SMDPP message, and sends it to the serving MSC for the mobile. The serving MSC translates the SMDPP message into an R-DATA message and relays it to the mobile. The mobile only provides a network layer acknowledgement (R-DATA ACCEPT or R-DATA REJECT) upon successful receipt of an OPTS message. This R-DATA ACCEPT or R-DATA REJECT is translated by the serving MSC into the smdpp message returned to the OTAF. The same TIA/EIA-41 procedures for locating, storing, and forwarding a CMT message apply to an OPTS message. TSAR may be applied to an OPTS message in order to send a message longer than what is supported by individual TIA/EIA-41 and TIA/EIA-136 messages. End-to-end encryption and separate message encryption across the air interface may be applied to an OPTS message to increase its security. An OPTS message may be carried over a DCCH or a DTC to a mobile. OPTS messages may also be carried across the S-BCCH as broadcast teleservice messages. Version information is typically added to a set of OPTS messages to ensure that a mobile does not overwrite newer programming with older programming.

Three primary OPTS messages are used to program a mobile for intelligent roaming. The IRDB Download and Extended IRDB Download messages program the IRDB, and the Alpha Tag Download message programs the alpha tags associated with the different service provider types. The Extended IRDB Download message is used when the size of the IRDB is greater than 255 octets. The Alpha Tag Download message may be sent with the IRDB Download or Extended IRDB Download message, or may be sent at a separate time. When a mobile receives an updated IRDB via OPTS, it overwrites its existing IRDB and performs a power-up scan (see Chapter 15) to confirm it is on the best service provider in the area. If a mobile rejects an OPTS message, the OTAF may resend the message depending on the reject cause returned by the mobile.

12.6.4 Generic UDP transport teleservice

The general UDP transport teleservice (GUTS) is an application data delivery service that supports the thin client architecture (TCA). As shown in Figure 12.8, TCA provides delivery of text-based information services from World Wide Web sites through the Internet and the cellular network to "smart" mobiles. The term *thin client* refers to the fact that the mobiles do not support the full graphical capabilities that a typical wired connection to the Internet provides due to the bandwidth limitations of the wireless environment. An application server may provide the gateway between the Internet and the cellular network.

GUTS is a two-way teleservice using the user datagram protocol (UDP) to identify the intended port where application data is to be delivered. One port assignment for a mobile supporting GUTS is a wireless application protocol (WAP) browser inside the mobile. The WAP

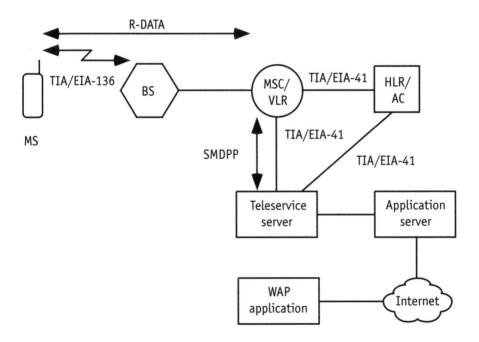

Figure 12.8 Thin client architecture used with GUTS.

browser allows the mobile user to access the Internet and private intranets from the mobile. WAP takes the place of the Hypertext Markup Language (HTML) and Hypertext Transport Protocol (HTTP) in TCA. Another port assignment for a mobile supporting GUTS is for text. The text can be treated similar to CMT by the mobile.

The same TIA/EIA-41 procedures for locating, storing, and forwarding a CMT message apply to a GUTS message. TSAR may be applied to a GUTS message in order to send a message longer than what is supported by individual TIA/EIA-41 and TIA/EIA-136 messages. End-to-end encryption and separate message encryption across the air interface may be applied to a GUTS message to increase its security. A GUTS message may be carried over a DCCH or a DTC to a mobile. GUTS messages may also be carried across the S-BCCH as broadcast teleservice messages.

There is currently only one GUTS message defined—the GUTS Data message. This message contains four information elements: Protocol Discriminator, Message Type, UDP Header, and User Data. The Protocol Discriminator, Message Type, and User Data are self-explanatory. The UDP Header identifies the source port and the destination port for the user data. For a mobile-terminated GUTS message, the source port is typically located in the application server and the destination port is located in the mobile. The ports are reversed for a mobile-originated GUTS message.

References

[1] Telecommunications Industry Association, *TIA/EIA Interim Standard: TDMA Cellular/PCS—Radio Interface—Mobile Station-Base Station Compatibility—Digital Control Channel, TIA/EIA/IS-136.1-A*, Oct. 1996, Section 7.

[2] Gallagher, M. D. and R. A. Snyder, *Mobile Telecommunications Networking with IS-41*, New York: McGraw-Hill, 1997, pp. 285–310.

Circuit-Switched Data Services

13.1 Introduction

Digital PCS offers both analog and digital circuit-switched data services. Analog circuit-switched data service is only available in the 800-MHz hyperband, while digital circuit-switched data service is available in both the 800-MHz and 1,900-MHz hyperbands. Key advantages of digital over analog circuit-switched data include increased capacity across the air interface, more reliable data calls, and data privacy via optional encryption. This chapter only highlights the analog circuit-switched data service available with Digital PCS, but describes in detail the digital circuit-switched data service. A packet data service for Digital PCS is described in Chapter 20.

13.2 Analog circuit-switched data service

Analog data service in Digital PCS is supported exactly as it is in AMPS [1,2]. Figure 13.1 shows the typical architecture for supporting analog

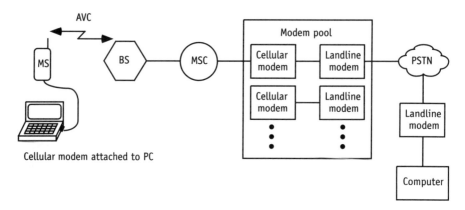

Figure 13.1 Typical architecture for analog circuit-switched data service.

data service. A cellular modem is used across the air interface to add extra error control and recovery not available in landline modems but necessary for the reliable transmission of data across the RF channel. Due to the overhead associated with providing this extra error control and recovery, cellular modems cannot offer the same usable data rates as landline modems. The serving MSC may contain a pool of cellular and landline modems to perform the necessary conversion between the modem protocols. There is not a single, ubiquitous analog cellular modem protocol, so the MSC typically contains different cellular modems for the different protocols. If the service provider does not offer this modem pool service, then the data user is on his or her own to guarantee that the same modem protocol is used at both ends of the data service.

When a mobile user has data to send from a PC, he or she connects the PC to the mobile through a cellular modem. A two-step dialing process is typically used, in which the user dials the access number for the modem pool prior to dialing the number of the destination computer. In the origination, the mobile requests an AVC for the circuit-switched data call. The call is set up to the modem pool by the serving MSC. When a cellular modem at the network answers, the user dials through to the destination computer and the data transfer begins. The modem pool may serve one or more MSCs. Billing for analog circuit-switched data calls may be the same as for voice calls, because there is no service code to notify the network to treat the calls differently. However, special billing for analog data calls is possible because the mobile user dials the modem pool first.

13.3 Digital circuit-switched data service

The digital circuit-switched data service for Digital PCS is also known as asynchronous data and group 3 fax service. The digital circuit-switched data service is carried across the air interface in a TIA/EIA-136 DTC. An uncompressed user data rate of 9.6 Kbps is achievable across a full-rate DTC, and operations in double- and triple-rate modes are defined to increase the uncompressed user data rate up to 28.8 Kbps. Cellular modems used for analog circuit-switched data service carry data in voice-band tones of varying frequency and amplitude. Vocoders, which are optimized to model human speech, do not model these voice-band tones to the required level of fidelity. For this reason, cellular modems used with analog circuit-switched data service cannot be used efficiently across a DTC. While this might appear to be a limitation, there is a corresponding advantage for digital because the modem functionality to convert data to voice-band tones and back again is not required. A new data service was developed for digitalDigital PCS to define the layer 2 and layer 3 protocols for digital circuit-switched data.

Digital circuit-switched data service uses the TIA/EIA-136 DTC layer 1 protocol described in Chapter 4. A Radio Link Protocol (RLP1) defined in IS-130 is used at layer 2 to improve the robustness of the data as it is transported over the RF channel. Layer 3 is provided by the IS-135 data part. The TIA/EIA-136 layer 3 described in Chapter 3 is used for data call setup. The protocol stack for digital circuit-switched data service is shown in Figure 13.2.

Figure 13.3 shows the key network elements for digital circuit-switched data. Referring to Figure 13.2 and 13.3, the mobile station (MS) is composed of data terminal equipment (DTE) and the mobile termination (MT). The DTE and the MT may be a single unit, or may be connected via a serial interface. An interworking function (IWF) replaces the analog modem pool for digital circuit-switched data service, and may be associated with one or many MSCs. It performs the protocol translation between the digital protocols carried over the air interface and the

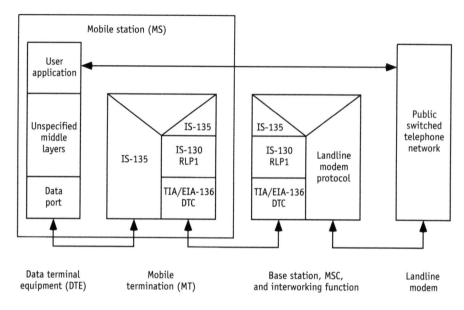

Figure 13.2 Protocol stack for digital circuit-switched data service.

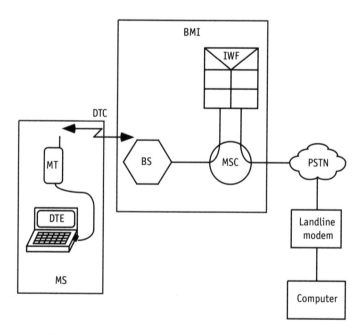

Figure 13.3 Network architecture for digital circuit-switched data service.

landline modem protocols used in the PSTN. The TIA/EIA-136 DTC physical layer is realized in the base station transceivers. Some portions of RLP1 may be realized in the base station transceivers, and other portions in the IWF. The base station, MSC, and IWF together form the BMI, as described in Chapter 7. In the context of IS-130, the IWF is called the BMI data communication equipment (DCE) because it performs the same functions as a conventional data or fax modem (also called DCE).

Figure 13.3 shows a computer as the landline end of the data connection. More generally, any DTE could be the end point for the data connection. This end point will be called the far-end DTE. Examples of far-end DTE include a fax machine or another mobile station. The figure also shows the IWF connected to the PSTN through the MSC. More generally, the IWF could connect directly to the PSTN, the integrated services digital network (ISDN), or a packet data network such as the Internet.

13.3.1 Radio Link Protocol 1

RLP1 provides link establishment, link supervision, acknowledged data transport, unacknowledged data transport, data qualification, data compression, data encryption, and flow control to layer 3 of the digital circuit-switched data service [3]. It requires a half-rate, full-rate, double-rate, or triple-rate TIA/EIA-136 DTC from layer 1. RLP1 supports two multiplexed data links, provided to layer 3 at two service access points (SAPs). SAP0 is used with IS-135 for transporting user data between the DTE and the far-end DTE. SAP1 is used with IS-135 for transporting command and control information, such as AT commands, between the DTE and the IWF.

Figure 13.4 shows a functional model of RLP1. RLP1 operates on octets of data called service data units (SDUs). A grouping of SDUs from the same SAP make up a protocol data unit (PDU). An RLP1 frame is constructed when PDUs from the same SAP are concatenated together and CRC bits (called the frame check sequence, FCS) are added. When this frame is encoded using convolutional encoding, it is known as an RLP1 encoded frame. RLP1 encoded frames are delivered to layer 1 for transmission and received from layer 1 for decoding and processing. An RLP1 frame is 27 octets (216 bits) in length, two or three octets of which are the

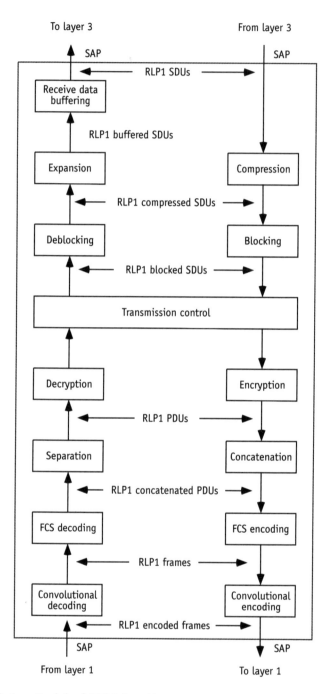

Figure 13.4 Model of RLP1 functions.

FCS, depending upon the level of CRC requested. RLP1 encoded frames are 260 bits in length. RLP1 processes SDUs, PDUs, and frames according to the functions listed in Table 13.1.

Table 13.1
Functions Performed by RLP1

RLP1 Function	Meaning
Receive data buffering	RLP1 provides flow control to layer 3. When the receive data buffer reaches an upper threshold, RLP1 stops acknowledging data, which causes the flow from the source to be stopped. When the receive data buffer reaches a lower threshold, RLP1 resumes acknowledgement to signal the source to begin sending data again. RLP1 buffers data independently for each SAP.
Compression and expansion	RLP1 provides compression of SDUs based on the V.42bis ITU standard. Compression is optional and may be turned on and off at layer 3.
Blocking and deblocking	RLP1 combines multiple compressed SDUs into larger data units (blocked SDUs). All of the compressed data units within a blocked data unit must be addressed to the same SAP. Flow control is active during the blocking operation to signal layer 3 to start and stop the data flow based on buffer thresholds.
Transmission control	RLP1 transmission control provides either an acknowledged mode or an unacknowledged mode of data transfer. In acknowledged mode, the receiver provides feedback indicating whether the data was received. It acknowledges received data and negative acknowledges missing data. Each SAP has its own transmission control. Unacknowledged operation is used when a constant data rate is required at the expense of a constant error rate.
Encryption and decryption	RLP1 performs encryption and decryption as a part of transmission control. The encryption mask is calculated in a different function, and XORed with the RLP1 PDUs prior to concatenation. Encryption is performed independently for each SAP and can be turned on or off based on instructions from layer 3.

Table 13.1 (continued)

RLP1 Function	Meaning
Concatenation and separation	RLP1 concatenates one or more PDUs from the same SAP after encryption and before frame check sequence encoding.
Frame check sequence encoding and decoding	RLP1 calculates a CRC for each concatenated PDU, and appends the CRC to the concatenated PDU. The CRC is used to determine if the received frame is in error. Any frame with an error is rejected.
Convolutional encoding and decoding	RLP1 adds channel coding to the frame of data in order to correct transmission errors across the air interface. A rate 1/2 convolutional code is used, then puncturing is applied to produce a net rate of 5/6.

The transmission control function is the most intricate part of RLP1. It has been designed to optimize throughput across the air interface for acknowledged data service [4]. There are two modes that may be used with acknowledged data service: basic mode and preemptive mode. In basic mode, a PDU is only retransmitted after the transmitter is sure it has not been received (a negative acknowledgement has been received). This makes the protocol very efficient because the ratio of the number of PDUs successfully received to the number of PDUs transmitted is kept high. Basic mode works well in good channel conditions, but in poor channel conditions retransmissions may stop completely because no feedback information is obtained. In this situation, preemptive mode can be used to retransmit unacknowledged and negatively acknowledged PDUs until they are acknowledged. Preemptive mode results in higher throughput and lower delays under poor channel conditions.

13.3.2 IS-135 data part

IS-135 provides the following services to the DTE user application: (1) call setup, supervision, and clearing; (2) AT command handling; (3) user data transport; and (4) on-line command signaling, break signaling, and signal leads. IS-135 resides in the MT and the IWF and interfaces to the DTE, RLP1, TIA/EIA-136, and the PSTN. It uses TIA/EIA-136 call

Circuit-Switched Data Services 239

control to perform the call setup, supervision, and clearing, at the request of either the user application within the DTE or the BMI. IS-135 assembles AT commands received from the DTE, forwards them to their destination (the MT, BMI, or both), and executes the commands.

Figure 13.5 shows a functional model of IS-135, which consists of the data port, the data port handler, the private channel, the connection manager, the DCE, and the command handler. The data port and the data port handler are only present in the MT implementation of IS-135, while the DCE is only present in the BMI implementation.

The MT communicates with the DTE through the data port. The data port autobauds to the correct setting for data rate, start bits, data bits, parity bits, and stop bits. It forwards these settings to the BMI DTE via the private channel. The data port handler routes command characters from the data port to the command handler and data characters from the data port to SAP0. It also performs flow control between the data port and SAP0. The command handler executes local and shared commands and

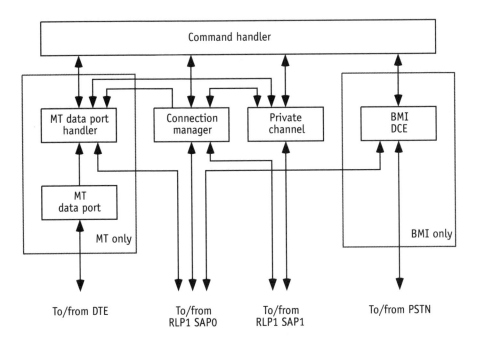

Figure 13.5   Model of IS-135 functions.

forwards shared and remote commands to the BMI via the private channel. The private channel messages are used to communicate between the MT and the IWF. These messages include commands, responses, and control information. The connection manager handles the RLP1 data link connections. The connection manager also uses the TIA/EIA-136 call control procedures to set up and terminate calls from a DCCH or ACC.

13.3.3 Data call processing

In order to originate an asynchronous data or group 3 fax call from the mobile to the PSTN, an AT command is entered from the DTE. The MT temporarily stores the AT command and originates a data call on the DCCH or ACC. The MT must include the desired service code (either asynchronous data or group 3 fax) in the origination. Based on the service code, the MSC assigns a DTC and connects the call to the appropriate port on the IWF. At this point, the logical links required by IS-130 are established and connected. The MT then issues the AT command that was stored. The dialed number from the AT command is used to connect the outgoing side of the call.

Three options are available to deliver data calls from the PSTN to the mobile: one directory number with prearrangement, one directory number per service, or two-stage dialing. Using one directory number with prearrangement, the cellular network will not know that the incoming call is a data call. The network assumes the call is a voice call and pages the mobile with a speech service code. The mobile would have to know ahead of time that the call is a data call in order to have the mobile configured to accept the call. This is possible with prearrangement, either by a voice call advising the user of an incoming data call or by prior agreement on a time for delivery of a data call. To configure the mobile to accept the data call, the user enters an AT command into the DTE, which would be sent to the MT as a local command. When the mobile receives the page for the data call, it requests an alternate service code in the page response. The MSC assigns a DTC and routes the call through the IWF to the mobile.

One directory number per service requires that multiple directory numbers be assigned to the mobile user—that is, a directory number for voice, a different one for asynchronous data, and yet another one for group 3 fax. Even though more than one directory number is assigned to the mobile user, the mobile only has one MIN. The best approach to mobile-terminated data calls may be two-stage dialing. With two-stage dialing, the service provider has dedicated numbers for each data service. The appropriate number is dialed first to indicate that this is a data call. The calling party waits for a secondary dial tone, then dials the subscriber's directory number. Because the MSC knows the data service type, it can page the mobile with the appropriate service code, assign a DTC, and route the call through the IWF.

References

[1] DeRose, J. F., *The Wireless Data Handbook*, Mendocino, CA: Quantum Publishing, 1994, pp. 31–34.

[2] Boucher, N. J., *The Cellular Radio Handbook*, 3d ed., Mill Valley, CA: Quantum Publishing, 1995, pp. 483–487.

[3] Telecommunications Industry Association, TIA/EIA-IS-130-A, *TDMA Wireless Systems—Radio Interface—Radio Link Protocol 1*, Sept. 1997.

[4] Nanda, S., R. Ejzak, and B. T. Doshi, "A Retransmission Scheme for Circuit-Mode Data on Wireless Links," *IEEE Journal on Selected Areas in Communications*, Vol. 12, No. 8, Oct. 1994, pp. 1338–1352.

Nonpublic Services

14.1 Network types

Wouldn't it be convenient if you were able to use your mobile phone as your primary phone while you were at home, in the office, or on the road? While you were at home, your mobile phone would operate like a cordless telephone, and all the usage charges would be associated with your residential service. While you were in the office, your mobile phone would operate like a wireless deskset (a service called business PCS [1] or wireless office), and your company would pick up the charges for business calls. While you were on the road, your mobile phone would be used like it is today. How about if you were able to use your mobile phone like an extension to your cordless telephone anywhere in the neighborhood in which you live, and not be tethered to your home by a few hundred feet? You could go to the swimming pool, a friend's house, or the neighborhood store and not have to worry about missing an important call. All of this is possible with Digital PCS through a feature called nonpublic services.

At the heart of nonpublic services is the concept of *network types*: public, private, and residential. Each cell site broadcasts information identifying the supported network types. Only mobile phones that subscribe to these network types may obtain service on the cell site. This means that network types can be used to create *closed user groups*. For example, a cell site covering an office building may broadcast that it only supports the private network type. All mobile phones that do not support this network type must ignore the cell site, while mobile phones that do support it may attempt to obtain service on the cell site. In essence, the cell site is closed to any mobile phones that do not support the private network type.

Why would anyone want to use network types to create closed user groups? The businesses in an office building may offer so much mobile phone usage that a cell site dedicated to the office building is warranted. Similarly, a residential base station could be used to offer cordless telephone service to a household, but not to the community at large. Special services, such as abbreviated dialing or distinctive ringing, could be provided to a closed user group on any of the network types.

The cell site identifies the network types it supports inside the Network Type information element within the System Identity message on the F-BCCH of the DCCH [2]. The Network Type information element is a bit map that identifies if public, private, and/or residential service is supported on the DCCH. Before the mobile phone can camp on the DCCH from power-up, it must read the System Identity message and determine from the Network Type information element if it can remain on the DCCH. This check is conducted by the mobile phone in the Service Aspects Determination procedure (see Chapter 8 for a more detailed description this procedure). If the DCCH does not support a network type that the mobile phone subscribes to, the mobile phone must mark the DCCH as ineligible and continue searching for another control channel.

The Network Type information element is also present in the Neighbor Cell and Neighbor Cell (Multihyperband) messages on the E-BCCH of the DCCH, and is populated for each neighbor list entry [3]. This means that the mobile phone is able to determine which network types are supported on all the neighboring control channels without having to synchronize to them and read the System Identity message. Any neighboring control channels marked with network types that the mobile phone does not subscribe to are deemed not to be viable neighbor list entries, and the mobile phone need not perform signal strength measurements on them during the Control Channel Reselection procedure (see Chapter 9 for details).

14.2 Private and residential system identities

The network type alone is not enough to define a closed user group. For example, suppose two cell sites offer service to two separate businesses. Both cell sites may identify themselves as supporting the private network type. There must be some way other than the network type to identify that the cell sites are offering service to different businesses. Otherwise, a mobile phone associated with one of them could obtain service on the cell site serving the other. *Private and residential system identities* are used in conjunction with network types to distinguish between closed user groups.

A private system identity (PSID) is used in conjunction with the private network type, while a residential system identity (RSID) is used in conjunction with the residential network type. Together, the PSID or RSID and network type define a closed user group. The system identity (SID) and system operator code (SOC) are used to identify a public network, but closed user groups are typically not associated with the public network type. When the user subscribes to a particular closed user group, their mobile phone is programmed with the network type and PSID or RSID that defines the closed user group. Then, when the mobile phone encounters a cell site matching the network type and PSID or RSID stored in memory, the mobile phone knows it may obtain service on the cell site. A mobile phone may be programmed with more than one PSID or RSID, thus allowing the user to subscribe to more than one closed user group; for example, a PSID for use at work and an RSID for use at home.

In order to inform the mobile phone user that service has been obtained on a particular closed user group, a name may be associated with each PSID or RSID: the PSID/RSID alphanumeric name, or alpha tag for short. Whenever the mobile phone is connected to a PSID or RSID on a cell site, it may display the appropriate alpha tag to the user, letting him or her know that specific services and billing arrangements apply in this area. If the mobile phone is using the public network type, it may display the alphanumeric SID broadcast by the cell site.

An example of how network types, PSIDs, and RSIDs can be used to form closed user groups is shown in Figure 14.1. One sector of the cell site is used to serve an office building. The sector broadcasts the fact that it supports the private network type with a PSID of 1. Another sector serves primarily mobile traffic, and broadcasts that it supports the public network type. The third sector serves both residential and public users, so it broadcasts that it supports the public and residential network types, with an RSID of 1 for residential users. Mobile phones connected to the public network type may display the alphanumeric SID "Digital PCS" to the user to identify the network. Mobile phones in use on the PSID or RSID may display a different alpha tag, associated with the broadcast PSID or RSID.

PSIDs and RSIDs are partitioned into ranges according to geographic limitations, as shown in Table 14.1. PSIDs fall into one of four ranges: SID-specific, SOC-specific, nationwide, or international [4]. All RSIDs are SOC-specific. For a SID-specific PSID, the mobile phone must store the SID, mobile country code, and PSID value, and must match these

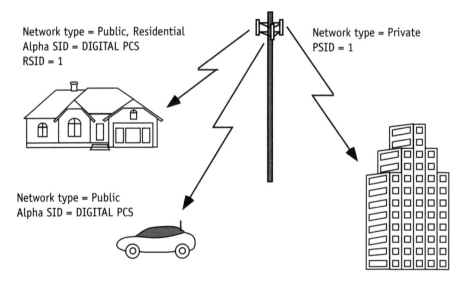

Figure 14.1 Forming closed user groups in Digital PCS.

Table 14.1
Ranges of PSIDs

PSID Range	Number in Range	Function
1–12,287	12,287	SID-specific PSIDs
12,288–53,247	40,960	SOC-specific PSIDs
53,248–57,347	41,000	Nationwide PSIDs
57,348–61,439	4,092	International PSIDs
61,440–65,535	4,096	Reserved

internally stored values with the broadcast values in order to declare a PSID match. This means that the same SID-specific PSID value may be used in different SID areas. Similarly, for a SOC-specific PSID (or any RSID), the mobile phone must store the SOC, mobile country code (for a national SOC), and PSID (or RSID) value, and must match these internally stored values with the broadcast values in order to declare a PSID (or RSID) match. If the SOC is in the international range, the mobile phone need not store and match the mobile country code. If the PSID is in the nationwide range, the mobile phone must store the mobile country code and PSID value. Finally, if the PSID is in the international range, the mobile phone only has to store and match the PSID value.

The cell site identifies the PSIDs and RSIDs it supports inside the PSID/RSID Set information element within the System Identity message on the F-BCCH of the DCCH. The PSID/RSID Set information element can identify up to 16 PSIDs or RSIDs that are supported by the cell site, although most implementations currently restrict the number to 5 per DCCH. Upon power-up, if the mobile phone identifies a DCCH broadcasting a PSID that matches one stored in the mobile phone, the mobile phone will attempt to register on that PSID. The mobile phone does this by sending a Registration message on the reverse access control channel (RACH) of the DCCH. Within the Registration message, the mobile phone will include the Selected PSID/RSID information element, which identifies to the network the PSID or RSID value the mobile phone is requesting service on. The network will typically check the subscriber profile for the mobile phone user and determine if the user subscribes to the selected PSID or RSID. If the mobile phone user subscribes to the PSID or RSID, the network will send the mobile phone a Registration Accept message on the SMS, paging, and access control channel (SPACH) of the DCCH. Otherwise, the network will send the mobile phone a Registration Reject message in a similar manner. Once the mobile phone has received a Registration Accept message on the selected PSID or RSID, the alpha tag associated with the PSID or RSID may be displayed to the user.

If the mobile phone reselects to a DCCH that is broadcasting a higher priority network type, PSID, or RSID, the mobile phone may attempt to register with that higher priority entity. The mobile phone sends a Registration message on the RACH of the new DCCH and includes the new PSID or RSID in the Selected PSID/RSID information element. The network may once again check the validity of the request before sending the mobile phone a Registration Accept or Registration Reject message.

The priorities of network types, PSIDs, and RSIDs may be stored in the mobile phone. Typically, an RSID would take priority over a PSID, and both would take priority over the public network type. Sophisticated rules may be used to identify the priorities of PSIDs and RSIDs if multiple entries are stored within the mobile phone for each network type, however. The priorities may be set by the mobile phone user through menu selections in the mobile phone. Alternately, the service provider may set the priorities by programming the mobile phone, either manually or over the air.

There are a number of methods for storing PSIDs and RSIDs within a mobile phone. The mobile phone's numeric assignment module (NAM) may be programmed with PSIDs and RSIDs manually or over the air using the over-the-air activation teleservice (see Chapter 12). Alternately, when the mobile phone registers with the network, the network may respond in the Registration Accept message with a list of PSIDs and RSIDs upon which the mobile phone is allowed service. Another method of acquiring PSIDs and RSIDs is for the mobile phone to send the network a Test Registration message on the RACH of the DCCH requesting service on a set of PSIDs and/or RSIDs that are broadcast on that DCCH. The network identifies in the Test Registration Response message sent on the SPACH the PSIDs and RSIDs on which the mobile phone is allowed to register. Thus, the mobile phone may have multiple storage areas for PSIDs and RSIDs. These storage areas are typically differentiated by the method that is used to acquire the PSIDs and RSIDs stored in the particular areas, either NAM, Registration Accept, or Test Registration.

The alpha tag associated with a particular PSID or RSID is stored in the mobile phone. It may be programmed into the NAM of the mobile phone or acquired through Test Registration. If Test Registration is used to acquire the alpha tag, the mobile phone stores the PSID/RSID alphanumeric name that is sent to it by the network in the Test Registration Response message. If the mobile phone acquires service on a PSID or RSID for which an alpha tag has not been stored, the mobile phone may display a default alpha tag to the user. The default alpha tag may be as simple as "PRIVATE" for a PSID, or "RESIDENTIAL" for an RSID.

14.3 Autonomous systems

When a system operates independent from the traditional cellular network but still provides services similar to and in the same frequency band with the cellular network, that system is called an *autonomous system* [5]. Autonomous system is a fancy name for a simple concept. Suppose a residential base station was used to offer cordless telephone service to a household but not to the community at large. The residential base station would probably include a scanner that would identify channels within the 800-MHz or 1,900-MHz frequency band that it could use without interfering with the cellular network. The residential base station would not be connected to the cellular network in the traditional sense in which a cell site is connected to the cellular network. That is, the cellular switch would not control the operation of the residential base station, and call handoff between the cellular network and the residential base station would not be provided. The residential base station would operate independent from, or autonomous of, the cellular network, and would therefore be classified as an autonomous system. A wireless business system could also operate as an autonomous system in a similar manner.

Autonomous systems provide advantages for both the service provider and the customer over traditional cellular service. Because an autonomous system can be initially programmed by the service provider and then left to operate on its own, the service provider does not need to continuously monitor and maintain the autonomous system. Unlike a cell site, the autonomous system is not tied to geographic location. This means that a customer may purchase and use an autonomous system in much the same way as he or she would a cordless telephone in the case of a residential base station, or a private branch exchange (PBX) in the case of a wireless business system. It should be noted, however, that the cellular frequencies used by an autonomous system in any specific geographic area are controlled by the service provider for that area, so the local service provider must authorize the use of all autonomous systems in its domain.

Digital PCS lends itself well to use with autonomous systems. First, a Digital PCS channel can fit within a small portion (30 kHz) of the frequency band. Second, the Digital PCS control channel can be located anywhere in the frequency band. Both of these attributes make it easier for an autonomous system to find channels it can reuse from the cellular network. Third, procedures are described in the Digital PCS standard to assist a mobile phone in finding an autonomous system. When a mobile phone does find an autonomous system, the network type and private and residential system identities described previously are used to provide nonpublic services to the mobile phone user.

An autonomous system may be composed of one or more cell sites, each broadcasting a DCCH. The DCCH frequencies of an autonomous system are known as private operating frequencies (POFs). A mobile phone that subscribes to an autonomous system stores the POFs for the autonomous system in its memory. Upon power-up, the mobile phone scans these stored POFs first, which guarantees that if the mobile phone is powered up in range of the autonomous system, it will find one of the autonomous system's control channels. If a DCCH is not found on one of the POFs, the mobile phone proceeds to scan for control channels in the usual manner (see Chapters 8 and 15). The mobile phone may store up to four POFs for each autonomous system.

The mobile phone must implement a new reselection trigger condition (see Chapter 9) to support reselection to the DCCH of an autonomous system. This is because autonomous system DCCHs are not present on the neighbor list of any DCCHs in the cellular network. (To make the autonomous system DCCHs appear on the neighbor list of the cellular network's DCCHs would require the service provider to keep track of the frequencies used by each autonomous system and continuously update neighbor lists, which could be an arduous task and make the neighbor lists extremely long.) This new reselection trigger condition is called a Priority System condition.

In order to declare a Priority System condition and reselect to an autonomous system DCCH, the mobile phone must have stored a set of public service profiles (PSPs) for the autonomous system. PSPs are the key characteristics of the cellular network control channels that overlay the autonomous system. For ACCs, the key PSP characteristics are the frequency and digital color code (DCC). For DCCHs, the key PSP characteristics are the frequency and digital verification color code (DVCC). The mobile phone may store up to four PSPs for each autonomous system.

Whenever the mobile phone is operating on a control channel in the cellular network that matches one of the stored PSPs, the mobile phone checks the stored POFs that correspond to the stored PSP. If the mobile phone determines that the POF is actually the desired autonomous system, it may reselect to the autonomous system and obtain service. Otherwise, the mobile phone remains on the cellular network control channel. This operation ensures that the mobile phone is not continuously scanning for the autonomous system, but intelligently monitors its surroundings to determine when it is in range of the autonomous system.

Figure 14.2 shows an example of an autonomous system with POFs and PSPs. In the figure, the POFs are denoted by the lowercase f (for frequency), while the PSPs are denoted by the uppercase F. The figure shows four POFs and four PSPs for the autonomous system that covers the building. The PSP may also include the SID of the cellular network. This will ensure that the mobile phone does not scan for the autonomous system when it is roaming in other SID areas.

POFs and PSPs are stored in the mobile phone automatically once the mobile phone camps on an autonomous system's DCCH. The mobile phone knows that it is camping on an autonomous system's DCCH by determining that the DCCH does not support the public network type, but only the private or residential network type. When the mobile phone reads the Neighbor Cell message on the E-BCCH of the autonomous system's DCCH, it stores the first four neighbor list entries marked with CELLTYPE of Regular or Preferred as the POFs (see Chapter 9 for a

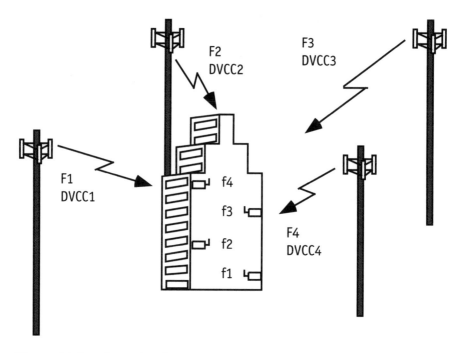

Figure 14.2 Example of an autonomous system.

description of reselection). Similarly, the mobile phone stores the first four neighbor list entries marked with CELLTYPE of Nonpreferred as the PSPs. Each time the mobile phone camps on the autonomous system's DCCH, it updates the POF and PSP entries for the autonomous system. In this manner, if any of the POF and PSP entries change as a result of the autonomous system scanning and determining that a different frequency or set of frequencies should be used, the mobile phone can track these changes. POF and PSP entries may also be updated manually or by using the over-the-air activation teleservice.

One other method of finding an autonomous system is provided for in Digital PCS. A new system search may be initiated by the mobile phone user in an attempt to acquire service on an autonomous system. This method is useful if POFs and PSPs have not been stored by the mobile phone or are stale. When the mobile phone user initiates a new system search, typically via a menu selection on the mobile phone, the mobile phone conducts a scan in an attempt to find autonomous system DCCHs.

If an autonomous system is found that is broadcasting a PSID or RSID matching one already stored in the mobile phone, the mobile phone displays the alpha tag for this autonomous system to the user. If the user selects the autonomous system, the mobile phone proceeds to obtain service on the autonomous system's DCCH and updates the POFs and PSPs based on the broadcast neighbor list as described above.

If in the course of conducting a new system search, the mobile phone finds the DCCH of an autonomous system that is broadcasting a PSID or RSID that does not match one stored in the mobile phone, the mobile phone may perform a Test Registration to determine whether it can obtain service on this autonomous system. If the Test Registration Response received by the mobile phone from the autonomous system indicates that it may obtain service with the selected PSID or RSID, the mobile phone displays the alpha tag associated with the new PSID or RSID to the user. If the user selects the autonomous system, the PSID or RSID is stored and the mobile phone proceeds to obtain service on the autonomous system's DCCH and stores the POFs and PSPs of the autonomous system.

References

[1] Clary, A., F. M. Marchetti, and S. Smith, "Business PCS—Trial Results," *IEEE Trans. Veh. Tech.*, Vol. 43, No. 3, Aug. 1994, pp. 722–726.

[2] Telecommunications Industry Association, *TIA/EIA Interim Standard: TDMA Cellular/PCS—Radio Interface—Mobile Station-Base Station Compatibility—Digital Control Channel, TIA/EIA/IS-136.1-A*, Oct. 1996, pp. 181, 263.

[3] Ibid., pp. 186, 190.

[4] Ibid., pp. 338–339.

[5] Ibid., pp. 169–173.

15

Special Considerations for 1,900-MHz Operation

15.1 Intelligent roaming

Intelligent roaming is a method for ensuring that a mobile phone obtains service from the best provider in an area without requiring user intervention. The "best" provider is typically either the home service provider or a roaming partner with which favorable airtime rates have been negotiated. If more than one roaming partner is available in an area, the best service provider may be the one that offers the mobile user the set of features and services that most closely matches those available in the home market. This is typically only possible from a service provider using the same air interface technology, such as Digital PCS, as the home service provider.

Before the 1,900-MHz frequency band was opened for cellular service, there were typically only two providers in an area—the 800-MHz A and B band providers. Most A-side service providers had roaming agreements with other A-side service providers, and the same was true for the B-side providers. Under most roaming scenarios, a mobile user was offered the best roaming rates if the mobile remained tuned to the home band—that is, a mobile homed to a B-side service provider would most likely desire to roam onto another B-side service provider. No user intervention was required to keep the mobile on the home band, so intelligent roaming was not necessary. Occasionally, an 800-MHz service provider would own both A- and B-side markets. Most of the time, any roaming between these markets was accomplished by requiring the user to manually switch the mobile to the alternate band through a system preference menu in the mobile. This had mixed results, however, as many mobile users either forgot how to switch bands or never learned in the first place. One might label this as "dumb" roaming.

The primary event that triggered the need for intelligent roaming was the opening of the 1,900-MHz frequency band for cellular service and the development of dual-band 800-MHz/1,900-MHz mobiles. It would be a daunting task to design a user-friendly system preference menu that included two 800-MHz bands and six 1,900-MHz bands. Furthermore, the coverage of preferred 1,900-MHz systems may not match those of 800-MHz systems for some time. A method for the mobile to automatically select the best band out of eight possible bands was needed. Enter intelligent roaming, which can also be used with single-band mobiles to automatically switch between the A- and B-sides for optimum service.

There are two basic ways of providing intelligent roaming to mobile users: network-directed and mobile-directed. In network-directed intelligent roaming, the mobile finds a control channel on one of the available bands and signaling is used to direct the mobile to the most appropriate band for it to obtain service on in the area. In mobile-directed intelligent roaming, the mobile contains all the information that is necessary to decide which band to operate on in any given area. Both air interface and intersystem signaling are required for network-directed intelligent roaming. There is also no guarantee that the serving system will support network-directed intelligent roaming, or that it will pass the intelligent roaming information to the mobile. For these reasons, mobile-directed intelligent roaming was chosen for Digital PCS.

With the mobile-directed method of intelligent roaming implemented for Digital PCS, the mobile stores information critical for roaming in what is called an intelligent roaming database (IRDB). Within the IRDB are lists of system IDs (SIDs) and system operator codes (SOCs) that identify service providers according to categories. The categories are home, partner, favored, neutral, and forbidden, in priority order. A mobile scans for control channels based on a band order, also stored in the IRDB, and searches for the highest priority service provider it can find that is not forbidden. If the mobile finds a control channel broadcasting an SID or SOC not in the IRDB, the service provider is classified as neutral. If the highest priority service provider that can be found is favored or neutral, the mobile will obtain service but occasionally scan for a higher priority service provider. The IRDB within a mobile may be updated as roaming agreement change using the over-the-air programming teleservice (OPTS) as described in Chapter 12. A flowchart describing the overall intelligent roaming process used by a mobile is shown in Figure 15.1. The intelligent roaming process is described in more detail below.

The home service provider must define the SID and SOC information stored in the mobile's IRDB. This is accomplished by categorizing service providers into home, partner, favored, neutral, and forbidden categories. The home service provider is determined by the home SID or home SOC stored in the mobile's NAM. If the home service provider has markets that use different SIDs and it is desired to provide home service in all of the markets, the home service provider may program the IRDB such that the SOC determines the home service provider. A partner service provider is the next highest category of roaming partners. A partner service provider is defined by a partner SID or partner SOC entry stored in the IRDB. A mobile quits scanning for service providers when it finds a home or partner service provider. For this reason, home and partner service providers are defined as *acceptable* service providers.

The next highest classification of service providers is the favored category. A favored service provider is defined by a favored SID or favored SOC entry stored in the IRDB. A mobile may obtain service on a favored service provider if an acceptable service provider is not available. A neutral service provider does not have a SID or SOC entry in the IRDB. A mobile may obtain service on a neutral service provider if an acceptable service provider is not available and a favored service provider is not either. Finally, a forbidden service provider is identified by a forbidden SID or SOC entry in the IRDB. A mobile should never obtain service on a

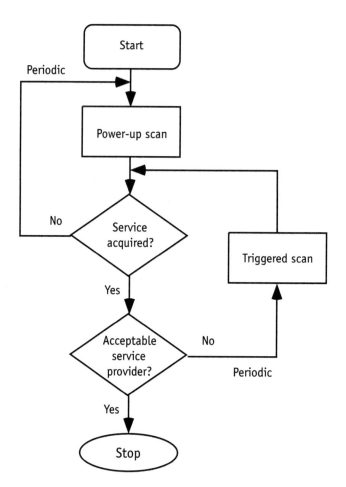

Figure 15.1 Flowchart of the overall intelligent roaming scanning process for a mobile.

forbidden service provider except to make an emergency call. A mobile has to exhaustively search the bands identified in the IRDB before obtaining service on a favored or neutral service provider. A mobile also conducts periodic scans of other bands at an interval defined in the IRDB in search of an acceptable service provider if it is in service on a favored or neutral service provider. For these reasons, favored, neutral, and forbidden service providers are defined as *unacceptable* service providers.

Special Considerations for 1,900-MHz Operation   259

The IRDB stores all the important information for the mobile to perform intelligent roaming. This includes the service provider information described previously and information that controls how the mobile is to scan for service. As summarized in Table 15.1, there are a number of parameters that control how the mobile is to scan. The primary purpose of the IR Control Data fields is to allow for different implementations of intelligent roaming by different service providers, thereby giving intelligent roaming the flexibility to meet wide-ranging business needs. The remainder of the parameters bound the mobile scanning process and will be described in greater detail as the different intelligent roaming scanning procedures are illustrated.

Table 15.1
IRDB Parameters Defining Mobile Scanning

IRDB Parameter	Meaning
IR Control Data—Home Only Enable	Defines whether to search only for the home service provider or for the highest priority service provider that is not forbidden.
IR Control Data—DHT Enable	Defines whether or not to use the DCCH history table during a power-up scan.
IR Control Data—Alpha Tag Enable	Defines whether to display the broadcast alphanumeric SID or an internally stored alpha tag when operating on a control channel.
IR Control Data—SOC Disable	Defines whether the SID or SOC takes priority when classifying a service provider.
IR Control Data—Enhanced DHT Enable	Defines whether or not to obtain service on a favored or neutral service provider based on a match in the DCCH history table.
IR Control Data—Triggered Scan Disable	Defines whether or not triggered scans are enabled when operating on a favored service provider.
IR Control Data—Nonpublic Priority Enable	Defines whether or not a mobile should obtain service on a private or residential system, regardless of the priority of the service provider's SID or SOC.

Table 15.1 (continued)

IRDB Parameter	Meaning
Band Order	Defines the order for the mobile to scan the 800-MHz and 1,900-MHz bands. If a frequency band is not in the band order, the mobile does not scan it except if an emergency call is placed.
NUM_CELLULAR	Defines the number of probability blocks to scan in an 800-MHz frequency band when searching for a DCCH.
NUM_PCS	Defines the number of sub-bands to scan in a 1,900-MHz frequency band when searching for a DCCH.
RESCAN_COUNT	With Rescan_Count_Initial_Value, defines the interval between triggered partial scans.
RESCAN_LOOP	Defines the number of times to perform a triggered partial scan before performing a triggered wideband scan.
History Threshold	Defines the signal strength below which a channel is ignored when searching for a DCCH during a historic search.
NUM_DHT	Defines the maximum number of entries to be stored in the DCCH history table.
NUM_BHT	Defines the maximum number of entries to be stored in the band history table.
Rescan_Count_Initial_Value	With RESCAN_COUNT, defines the interval between triggered partial scans.
Sub-band Priority Order	Defines the order in which sub-bands are scanned within any 1,900-MHz frequency band.

A power-up scan is performed as shown in the flowchart of Figure 15.2. A mobile first conducts what is called a *historic search* of a list of channels where it either has obtained service in the past or has obtained information in the past indicating there is a high probability of obtaining service on these channels. There are two parts to the historic search. The first part is a check of private operating frequencies (POFs) for nonpublic service, as described in Chapter 14. If no service is found on the POFs, the

Special Considerations for 1,900-MHz Operation

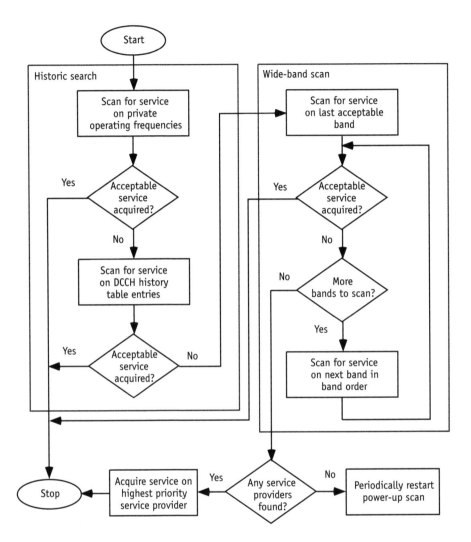

Figure 15.2 Flowchart of the power-up scan.

entries in the DCCH history table (DHT) are checked. The received signal strength on the POFs and DHT entries must be above HISTORY_THRESHOLD in order to be considered during the historic search. If no service is found on any of the DHT entries, the mobile performs a wideband scan.

The IR Control Data entries in the IRDB can modify the outcome of the historic search portion of the power-up scan. The DHT Enable flag must be set for the mobile to check any of the DHT entries. If the Enhanced DHT Enable flag is set, the mobile can obtain service on a favored or neutral service provider during the historic search. If the Nonpublic Priority Enable flag is set, the mobile can override the service provider priorities if a matching private or residential system ID is found during the historic search. If neither the Enhanced DHT Enable nor the Nonpublic Priority Enable flags are set, the mobile can only obtain service on a home or partner service provider during the historic search. If the SOC Disable flag is set, the service provider category is determined by the priority of the SID; otherwise, the category is determined by the highest priority of the SID or SOC.

If the mobile completes the historic search portion of the power-up scan without obtaining service, it enters a wideband scan. During a wideband scan, the mobile scans the 800-MHz and 1,900-MHz frequency bands in an ordered sequence trying to obtain service from the highest priority service provider that is available. When a wideband scan is entered from a power-up scan, the first frequency band that is checked is the last band where the mobile previously found a home or partner service provider. This guarantees that if the mobile is powered up in the same general area where it was powered down, and if that area was served by an acceptable service provider, the mobile will quickly find service on that same service provider.

If the mobile fails to find service on an acceptable service provider while checking the last acceptable band, the mobile begins searching the remaining bands in their order of appearance in the IRDB band order. The mobile only searches those bands that are listed in the band order, and obtains service immediately when an acceptable service provider is found. If an unacceptable service provider is found on a band during a wideband scan, the control channel and service provider information are temporarily saved by the mobile station.

If the mobile searches all the bands in the band order without finding a control channel of an acceptable service provider, it returns with a list of the unacceptable service providers it has found. If the mobile is performing a power-up scan, the mobile obtains service on the highest priority

Special Considerations for 1,900-MHz Operation 263

unacceptable service provider it found during the wideband scan. The mobile will only obtain service on a forbidden service provider if the mobile user desires to make an emergency call, however. In the event that no service is available, the mobile will periodically restart the power-up scan. As with the historic search, if the Nonpublic Priority Enable flag is set, the mobile may obtain service immediately if a matching private or residential system ID is found, regardless of the service provider category.

Procedures are defined within intelligent roaming for scanning within the 800-MHz and 1,900-MHz frequency bands. These procedures are invoked by the mobile during a wideband scan and during a triggered scan, which is described later in this chapter. There are different procedures for scanning in the 800-MHz and 1,900-MHz frequency bands, primarily because ACCs may be present at 800 MHz, but are not present at 1,900 MHz.

The 800-MHz scanning procedure is illustrated in Figure 15.3. The mobile first attempts to find an ACC in a similar manner as an AMPS mobile would (see Chapter 1). If an ACC is found, the mobile determines whether this system is capable of digital operation. It does this by checking the protocol capability indicator (PCI) bit in the overhead message train of the ACC. If the system is capable of digital operation, the mobile waits up to five seconds to determine if the Control Channel Information message is present directing the mobile to a DCCH. If the system is not capable of digital operation or the mobile does not find the Control Channel Information message on the ACC, it proceeds to read the SID from the ACC and make a service provider determination based only on this information.

If the mobile reads a Control Channel Information message on the DCCH, it attempts to tune to the DCCH and read service provider information there. This is necessary because the SOC and any private or residential system IDs supported by the system are only present on the DCCH. If the signal strength on the DCCH is too low or the mobile cannot synchronize to the DCCH, it scans the probability block where the DCCH is located for a better DCCH. If the mobile is able to synchronize to a DCCH, it reads the SID, SOC, and any private and residential system IDs and makes a service provider determination. If the mobile cannot find

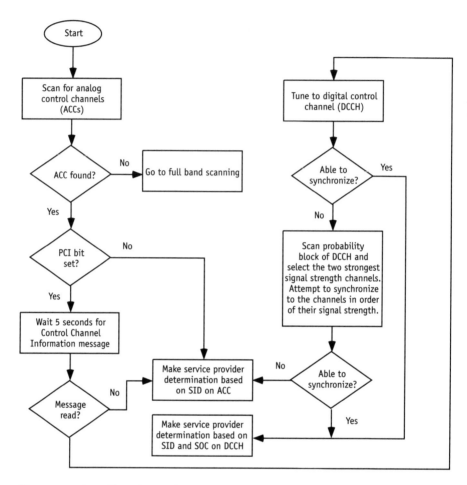

Figure 15.3 Flowchart of 800-MHz scanning.

a DCCH, it returns to the ACC and makes a service provider determination as described previously. In the even that no control channels can be found, the mobile executes a full-band scan.

A full-band scan is executed by a mobile when no control channels have been found during an 800-MHz scan or a 1,900-MHz band is being scanned. The full-band scan is illustrated in the flowchart of Figure 15.4. The mobile scans the highest NUM_CELLULAR 800-MHz probability blocks or NUM_PCS 1,900-MHz sub-bands searching for a DCCH. If a DTC is found while searching for a DCCH, the mobile checks for DCCH

Special Considerations for 1,900-MHz Operation 265

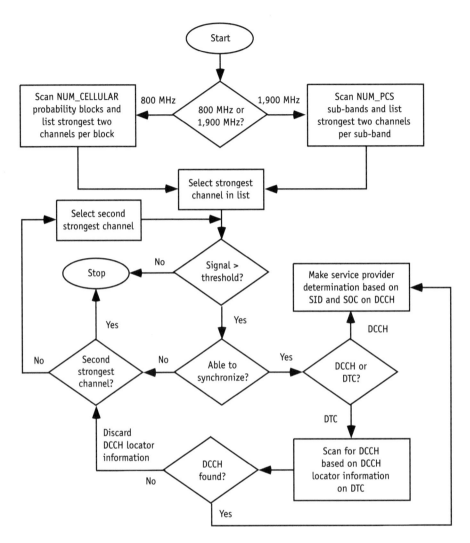

Figure 15.4 Flowchart of full-band scanning.

locator information on the DTC and follows it to the DCCH. If a DCCH is found during a full-band scan, the mobile makes a service provider determination based on the information it reads.

If a mobile obtains service on an unacceptable service provider, it will periodically scan for an acceptable service provider by executing a triggered scan, as shown in the flowchart of Figure 15.5. The IRDB

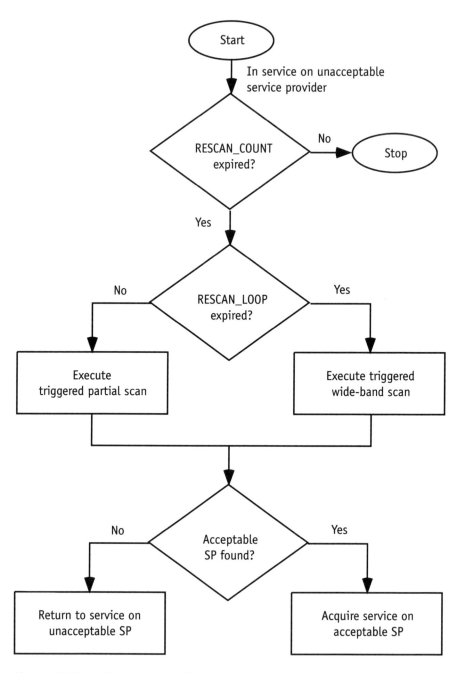

Figure 15.5 Flowchart of triggered scanning.

parameter RESCAN_COUNT (as modified by Rescan_Count_Initial_Value) determines the interval between triggered scans. There are two types of triggered scans—a triggered partial scan and a triggered wideband scan. A triggered partial scan is a scan of a single band, while the triggered wideband scan is a scan of all the bands in the IRDB band order (except for the current band, of course). A number of triggered partial scans are executed before a triggered wideband scan is performed. The IRDB parameter RESCAN_LOOP determines the number of triggered partial scans performed before executing a triggered wideband scan.

The band scanned in a triggered partial scan is either the last band where an acceptable service provider was found or the band identified in the band history table (BHT) where a higher priority service provider has been found in the past due to a triggered scan. The procedure the mobile uses to scan the band in a triggered partial scan is one of the band-scanning procedures previously described. The procedure the mobile uses to scan multiple bands in a triggered wideband scan is the wideband scanning procedure, also described previously. If the mobile does not find an acceptable service provider as a result of a triggered scan, it returns to the control channel of the current unacceptable service provider. The Triggered Scan Disable field in the IRDB can turn off triggered scanning from a favored service provider. The mobile executes a triggered scan in such a way that the probability of missing a page on the current system is minimized.

The IRDB in a mobile can be updated using the over-the-air programming teleservice (OPTS) described in Chapter 12. This may be accomplished while the mobile is operating either on a DCCH or a DTC. Immediately following the download of a new IRDB, the mobile performs a power-up scan. Reprogramming of the IRDB is a valuable feature for the home service provider, because roaming agreements may change in the useful lifetime of a mobile.

15.2 Hyperband reselection and handoff

Functionality is provided within Digital PCS to direct a mobile from the 800-MHz band to the 1,900-MHz band, and vice versa, while camping on a DCCH or in conversation on a DTC. This functionality is known as hyperband reselection and handoff. Hyperband reselection and handoff allows for seamless operation between Digital PCS systems on different frequency bands, or hyperbands. The systems may be operated by the same service provider or different service providers. The mobile user does not necessarily know that the mobile has changed hyperbands, but only that service is still offered. To take advantage of hyperband reselection and handoff, the mobile must be capable of dual-band operation.

Hyperband reselection and handoff are necessary to provide seamless service in areas where the 800-MHz coverage is more extensive than the 1,900-MHz coverage. Without this functionality, a mobile operating on a 1,900-MHz Digital PCS system would experience a no-service condition for some time after leaving the coverage of that system. During the no-service condition, the mobile may miss a page for an incoming call, not allow the origination of a call, or drop the current call. Without hyperband reselection and handoff, the mobile would need to execute the intelligent roaming procedures described previously every time it left the coverage area of the 1,900-MHz system.

Hyperband reselection is accomplished in much the same way as described in Chapter 9. The DCCHs near the border between 1,900-MHz and 800-MHz coverage include a layer 3 message called the Neighbor Cell (Multihyperband) message on the E-BCCH. This message can contain information on neighboring control channels in the current hyperband and the other hyperband. If the Neighbor Cell message and the Neighbor Cell (Multihyperband) message are both included on the DCCH, a mobile capable of operation on the other hyperband considers the Neighbor Cell (Multihyperband) message to take precedence. The mobile will read the message and reselect to a control channel in the other hyperband when the appropriate trigger condition occurs, as described in Chapter 9.

It is possible for a hyperband reselection to occur from a 1,900-MHz Digital PCS system to an 800-MHz AMPS-capable system that does not support Digital PCS. In this case, the Neighbor Cell (Multihyperband) message on the 1,900-MHz DCCH has neighbor list entries for the 800-MHz ACCs. This allows for seamless downward reselection from

1,900 MHz to 800 MHz, but does not allow reselection in the other direction. A mobile operating on the 800-MHz AMPS system and desiring service on the 1,900-MHz Digital PCS system must rely on intelligent roaming triggered scans to obtain service on the 1,900-MHz system.

Hyperband handoff occurs when a mobile is operating on a DTC and is handed off by the serving system to a DTC or AVC on the other hyperband. In border cells between 1,900-MHz and 800-MHz coverage, a base station will send a dual-band mobile a Hyperband Measurement Order on the FACCH of the DTC that includes channel and hyperband information it desires to receive channel quality measurements on from the mobile. The mobile acknowledges the Hyperband Measurement Order, measures the received signal strength on the desired channels during inactive periods on the DTC, and reports them to the serving base station in a Channel Quality Measurement message on the FACCH or SACCH. The serving Digital PCS system uses the channel quality measurements to determine the best time to hand off the mobile to the candidate system on the other hyperband.

Intersystem signaling may be used to request handoff measurements from the candidate system. These handoff measurements provide additional validation that a handoff to the candidate system is appropriate. However, the candidate system is operating on a different hyperband and may not have resources to perform a received signal strength measurement on the mobile's transmissions. When the serving system determines a handoff to the candidate system on the other hyperband should occur, intersystem signaling is used to request facilities to accommodate the handoff from the candidate system. When the candidate system notifies the serving system that the handoff will be accepted, the serving system sends the mobile one of two handoff messages on the FACCH. If the handoff is to a DTC, the Dedicated DTC Handoff message is sent to the mobile. If the handoff is to an AVC, the Handoff message is sent instead.

Hyperband handoff is not supported from an AVC, which is only present at 800 MHz, to a 1,900-MHz DTC. At the end of a call on an AVC, hyperband reselection or an intelligent roaming is required to obtain service on a 1,900-MHz system. It is also possible to send information on call release, directing a mobile to a DCCH in another hyperband.

15.3 Dual-band, dual-mode mobiles

A dual-band mobile is capable of operating in both the 800-MHz and 1,900-MHz hyperbands. A dual-mode mobile is capable of both digital and analog operation. All Digital PCS mobiles are dual-mode, but not all are dual-band. Hardware and minor software differences exist between single-band 800-MHz mobiles and dual-band mobiles.

Figure 15.6 shows a simple block diagram of a dual-band, dual-mode mobile, focusing on the RF portion of the mobile. Extra hardware required for a dual-band mobile is labeled in the figure. It is assumed that a broadband low-noise amplifier (LNA) covering 800 MHz and 1,900 MHz is used for both single-band and dual-band mobiles. A 1,900-MHz power amplifier (PA), 1,900-MHz bandpass filters (BPFs), a voltage-controlled oscillator (VCO), a mixer, and RF switches represent the majority of the additional hardware required for a dual-band mobile. Note that a duplexer is not required at 1,900 MHz unless multirate operation is supported—that is, double- or triple-rate circuit-switched data. A duplexer is required at 800 MHz to support AMPS operation in which transmission and reception occurs at the same time. For full-rate operation on the DCCH and the DTC, however, mobile transmissions and receptions are offset in time, so a duplexer is not required.

The software differences between a dual-band and single-band Digital PCS mobile are minimal. The major new piece of software required for a dual-band mobile is the layer 3 software to support the intelligent roaming procedures, hyperband reselection, and hyperband handoff. However, all single-band Digital PCS mobiles in the future will also support intelligent roaming between the two 800-MHz bands.

Special Considerations for 1,900-MHz Operation

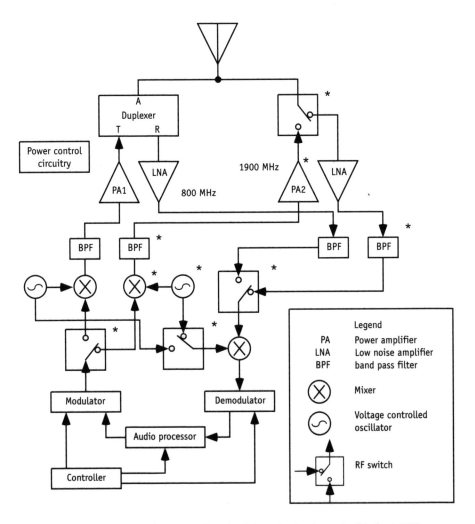

Figure 15.6 Block diagram of a dual-band, dual-mode Digital PCS mobile.

Authentication, Privacy, and Encryption

16.1 CAVE

The cellular authentication and voice encryption (CAVE) algorithm is at the heart of authentication, privacy, and encryption for Digital PCS. Therefore, it is appropriate to begin this chapter with a description of CAVE. However, disclosure of the CAVE algorithm is subject to the export jurisdiction of the U.S. Department of State as specified in the International Traffic in Arms Regulations (ITAR). Therefore, only those aspects of CAVE that are publicly available will be described. It is ironic, however, that the assessment of any cryptanalysis attack against a cryptographic algorithm such as CAVE begins with the assumption that the algorithm is known to the attacker.

The CAVE algorithm is used to calculate shared secret data (SSD), authentication responses, voice privacy masks, and signaling message encryption keys. It is a randomization algorithm that can be implemented in software. Due to its mathematical properties, knowledge of the CAVE algorithm combined with data sent over the air interface is not sufficient for an attacker to derive the A-Key, SSD, or correct authentication responses. For example, an exhaustive search through all possible values of SSD looking for the value that produces the known authentication response would require an impractical amount of computation and would not yield a unique solution. A unique solution would result if enough authentication responses were captured, but a large number of these responses would be required in order to converge on the correct SSD. Another attack to find the SSD would be to run the CAVE algorithm in reverse with the known authentication responses, but this is not practical since many final states are possible.

The inputs to the CAVE algorithm differ based on the procedure in use. For the generation of an authentication response, the inputs are a portion of the SSD, a portion or all of the mobile identification number (MIN), the electronic serial number (ESN) of the mobile, and a random number. For the updating of SSD, the inputs are the A-Key of the mobile, the authentication algorithm version (AAV), the ESN of the mobile, and a random number. For generation of the voice privacy mask and signaling message encryption key, inputs similar to those used to generate the authentication response are used. Typically, the CAVE algorithm is run in the mobile and the authentication center (AC, see Chapter 7). The remainder of this chapter describes the Digital PCS authentication, privacy, and encryption procedures that use the CAVE algorithm.

16.2 A-Key and SSD

The A-Key is a 64-bit secret key known only to the mobile and the home AC [1]. It is stored in the mobile's permanent security and identification memory, and is not viewable but is changeable. It is never transmitted over the air. The A-Key is used by the mobile and the home AC to generate the SSD, which is used in the Digital PCS authentication, privacy, and encryption algorithms.

The mobile may come to the service provider from the manufacturer with the A-Key either preprogrammed or default (set to all zeros). If it is preprogrammed, the service provider obtains the A-Key and ESN pair securely from the manufacturer for storage in the AC. The secure method of A-Key transfer is typically either via electronic data interchange (EDI) or an encrypted diskette shipped through registered mail. The A-Key may be programmed into the mobile manually. When the A-Key is entered into the mobile manually, a checksum is also entered to ensure the appropriate A-Key is stored. The A-Key may also be generated in the network and in the mobile through the over-the-air activation teleservice (see Chapter 12).

The SSD is a 128-bit pattern stored in the mobile's semipermanent memory and in the home AC. The first 64 bits of the SSD, called SSD-A, are used to support the authentication procedures. The second 64 bits of the SSD, called SSD-B, are used to support the voice privacy and message encryption procedures. The SSD is generated by the CAVE algorithm using the mobile's A-Key and ESN, plus a random number and the authentication algorithm version. The SSD may be updated in the mobile and the home AC through network and air interface signaling. In the network, the AC may or may not share the SSD with the serving system. If the SSD is not shared, all authentication processing requires TIA/EIA-41 signaling between the serving system and the home system. Keeping the SSD local to the home AC is more secure, however.

16.3 Authentication procedures

Authentication is the process of verifying the authenticity of a mobile. It was developed to prevent cloning fraud. Prior to authentication, a mobile's ESN and MIN could be obtained by an attacker and placed into another mobile in a process called cloning. A cloned mobile could be used to illegally obtain service and fraudulently make calls without paying for service. Cloning fraud represents lost revenue for the service provider because it uses RF and network resources that would otherwise be used by paying customers. Cloning was relatively easy because a mobile's ESN and MIN could be read off the air with a relatively inexpensive receiver. Authentication prevents cloning fraud because the fraudster is required to also know the value of SSD in the mobile, and the SSD is never transmitted over the air. If the SSD is compromised, it can be updated by the network because the SSD is generated from a secret key (A-Key) known only to the home AC and the mobile.

There are two general types of authentication: global challenge and unique challenge [2,3]. In global challenge authentication, every mobile is directed through broadcast information to send an Authentication message to the network coincidental with certain other messages. In unique challenge authentication, a mobile is directed through point-to-point signaling to send an Authentication message to the network. Global challenge is required for voice privacy and message encryption to be activated. Its primary drawback is the signaling overhead required for every mobile.

In global challenge authentication, a field that is broadcast on both the DCCH and the ACC, called the AUTH bit, is set to 1 to indicate that a mobile should authenticate. A Digital PCS system may also broadcast the Auth Map information element on the DCCH to indicate exactly which messages the mobile is required to send an Authentication message coincident with. If the Auth Map is not broadcast, the mobile is required to send an Authentication message coincident with all of the following messages: Registration, Origination, Page Response, R-DATA, and SPACH Confirmation. A 32-bit random number, called RAND, is also broadcast on the DCCH and ACC and is periodically changed by the network. The mobile uses SSD-A, RAND, its ESN, and other variables in the CAVE algorithm to calculate an 18-bit authentication response (AUTHR), which it sends to the network in an authentication message. The mobile is considered authentic if the AUTHR value calculated by the network matches the AUTHR value calculated by the mobile. This confirms that the mobile contains the correct SSD-A, ESN, and other parameters.

The global challenge authentication process is depicted in Figure 16.1. The serving network broadcasts RAND in the Access Parameters message on the F-BCCH of the DCCH, and in the Random Challenge A and Random Challenge B global action messages on the overhead message train of the ACC. At the time of mobile access on the reverse control channel, the mobile executes the CAVE algorithm using the following as input: RAND, SSD-A, the mobile's ESN, and either the last seven digits of the mobile's MIN (called MIN1) or the dialed digits (only for the case of an authentication at the time of call origination). The AUTHR value calculated by the mobile through execution of the CAVE algorithm is sent

Authentication, Privacy, and Encryption   277

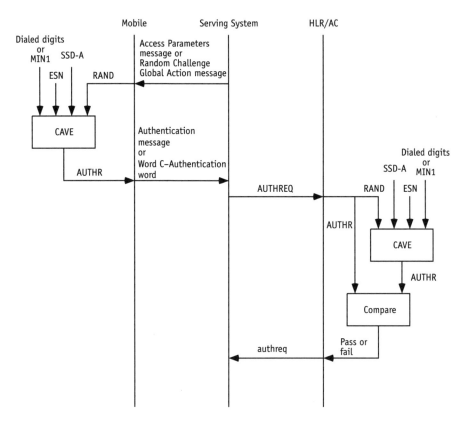

Figure 16.1   Global challenge authentication process.

to the network in either the Authentication message on the DCCH or Word C–Authentication Word on the ACC.

The authentication of the mobile's response to the global challenge takes place differently depending upon whether or not the mobile's SSD is shared between the home network and the serving network. Figure 16.1 shows the case for which the SSD is not shared. When AUTHR is received by the serving network, the serving network formulates a TIA/EIA-41 Authentication Request Invoke (AUTHREQ) message and sends it to the mobile's HLR/AC. The AUTHREQ includes the value of RAND used by the mobile and the AUTHR value calculated by the mobile. The AC executes the CAVE algorithm with the same inputs as the mobile, calculates AUTHR, and compares its calculation with that of the mobile. If the authentication results match, the mobile is considered authentic. Otherwise, the mobile is considered fraudulent and service is denied. The HLR/AC formulates an TIA/EIA-41 Authentication Request Return Result (authreq) message and sends it to the serving network to indicate if the mobile passed or failed authentication. For the case of shared SSD, the serving network performs the authentication of the mobile instead of the home network.

In unique challenge authentication, shown in Figure 16.2, the HLR/AC initiates the authentication process. It may do so for any number of reasons, including a failed global challenge, a successful global challenge in a high fraud area, or periodically. Unique challenge is a more

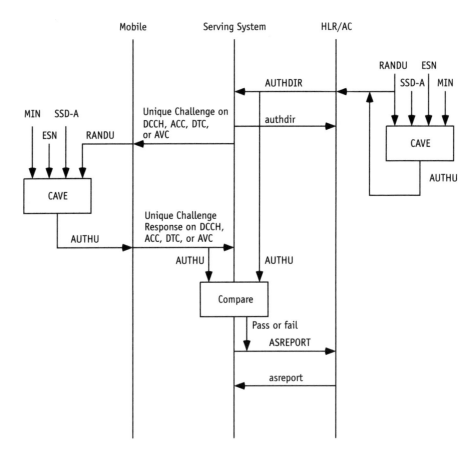

Figure 16.2 Unique challenge authentication process.

secure means of authentication than global challenge because the random number used in the challenge, called RANDU, is different for every challenge. If a fraudster has cracked the authentication response for a global challenge because RAND is broadcast and does not change rapidly, then unique challenge can be used to defeat this attack.

The AC typically initiates a unique challenge by generating a value of RANDU and executing the CAVE algorithm with this random number and the mobile's SSD-A, ESN, and MIN as input. The resulting CAVE output is an 18-bit unique challenge authentication response (AUTHU) that the AC temporarily stores. The HLR/AC then formulates a TIA/EIA-41 Authentication Directive Invoke (AUTHDIR) message, including RANDU and AUTHU, and sends it to the serving network. The serving network may send the unique challenge to the mobile on a DCCH, ACC, DTC, or AVC. Once the mobile receives the unique challenge, it executes the CAVE algorithm with the same inputs as used by the AC and sends the calculated value of AUTHU back to the serving network. The serving network compares the AUTHU supplied by the HLR/AC to that calculated by the mobile. If there is a match, the mobile is considered authentic. Otherwise, the mobile is considered fraudulent and service is denied. The serving network notifies the HLR/AC of the result of the unique challenge authentication by formulating a TIA/EIA-41 Authentication Status Report Invoke (ASREPORT) message and sending it to the mobile's HLR/AC.

The preceding description of the unique challenge authentication process assumes that SSD is not shared between the mobile's HLR/AC and the serving network. If SSD is shared, the serving network may choose to initiate a unique challenge at any time. For this case, the serving network instead of the mobile's HLR/AC generates RANDU and executes the CAVE algorithm to calculate AUTHU.

Another parameter sometimes used in the authentication process is the call history count (COUNT). COUNT is stored in the mobile and in the AC, and can be used to detect full clones. A full clone is a mobile that has obtained the ESN, MIN, A-Key, and SSD of a legitimate mobile. COUNT is incremented in a mobile only upon receipt of a Parameter Update message from the network. The current value of COUNT is sent to the network as part of a mobile's authentication response. If the value of COUNT reported by the mobile does not match that stored in the AC, the mobile may be denied service. The mobile does not challenge the network upon receipt of a Parameter Update message, however, so there is the potential for a network impersonator to fraudulently update the value of COUNT in a mobile. Care must therefore be exercised in the use of COUNT as a weapon against fraud.

16.4 Updating shared secret data

The SSD is an important aspect of the global and unique challenge processes, as it is the key piece of information that allows the authentic mobile to calculate the correct authentication response and the fraudulent mobile to calculate an incorrect response. The AC may elect to update the SSD in a mobile if it is suspected the SSD has been compromised. It may also routinely update a mobile's SSD upon return to the home network after the mobile has roamed into another network where the SSD has been shared. The mobile always challenges the network when an SSD update is initiated in order to keep a network impersonator from arbitrarily updating a mobile's SSD and wreaking havoc on the legitimate network.

The SSD update process is shown in Figure 16.3. The AC begins the process by executing the CAVE algorithm using as input the mobile's A-Key, ESN, and a random number called the SSD random variable (RANDSSD). The output of the CAVE algorithm in this case is the new value of SSD in two parts, called SSD-A_New and SSD-B_New. The HLR/AC formulates a TIA/EIA-41 Authentication Directive Invoke (AUTHDIR) message, including the RANDSSD value, and sends it to the serving network. The serving network may send the mobile an SSD Update Order message on the DCCH, ACC, DTC, or AVC. Upon receipt of the SSD Update Order, the mobile executes the CAVE algorithm with the same inputs as used by the AC and calculates SSD-A_New and SSD-B_New. The mobile then formulates a Base Station Challenge Order message that includes a new random number called the base station random variable (RANDBS) and sends it to the serving network over the air interface. The serving network relays RANDBS to the mobile's HLR/AC in a TIA/EIA-41 Base Station Challenge Invoke (BSCHALL) message.

Authentication, Privacy, and Encryption

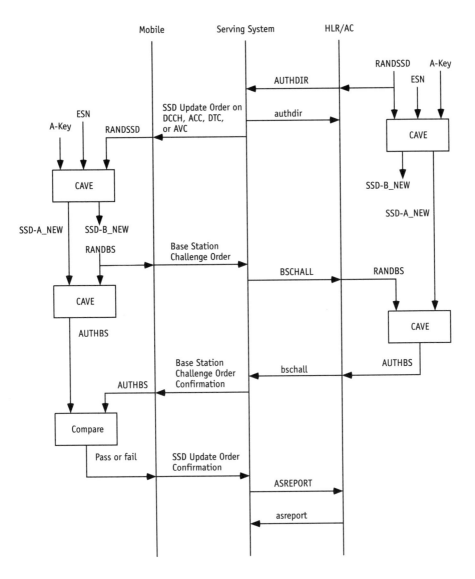

Figure 16.3 SSD updating process.

Both the mobile and the HLR/AC execute the CAVE algorithm again using SSD-A_NEW and RANDBS as inputs. The result is an 18-bit base station authentication response (AUTHBS). The HLR/AC returns its calculation of AUTHBS to the serving network in a TIA/EIA-41 Base

Station Challenge Return Result (bschall) message. The serving network relays AUTHBS to the mobile in a Base Station Challenge Order Confirmation message over the air interface. The mobile compares the value of AUTHBS it calculated with that calculated by the AC. If the values match, the authenticity of the network has been guaranteed and the mobile replaces its previous SSD with the updated SSD. Otherwise, the mobile discards the updated SSD and retains its previous SSD. The mobile formulates a SSD Update Order Confirmation message and sends it to the serving network indicating success or failure of the SSD update process. The serving network relays the SSD update result to the HLR/AC via the TIA/EIA-41 Authentication Status Report Invoke (ASREPORT) message.

The preceding description of the SSD update process assumes that SSD is not shared between the mobile's HLR/AC and the serving network. If SSD is shared, the HLR/AC sends the new SSD and RANDSSD to the serving network and the serving network processes the Base Station Challenge. Once the serving network receives the SSD Update Order Confirmation, an ASREPORT message is sent to the HLR/AC to indicate success or failure of the SSD update process.

16.5 Voice privacy and signaling message encryption

Voice privacy provides an enhanced level of privacy to the already digitized voice transmitted over the Digital PCS air interface. It is only available during the conversation state on a DTC. Voice privacy may be activated at the initiation of a call or anytime during a call. A mobile may only request that voice privacy be provided; the network is always in control of activating voice privacy. A menu option in the mobile's user interface is typically provided to allow the user to turn on or off voice privacy. Most mobiles provide an audible indication at the time of call establishment or handoff if voice privacy was requested but is not provided.

Voice privacy is provided through the generation and application of a mask, called the voice privacy mask (VPMASK), which is used with the digitized voice to provide an encrypted signal for transmission over the air interface. The voice is decrypted at the receiving end by applying the VPMASK to the received encrypted signal. The VPMASK is generated by executing the CAVE algorithm using as input the mobile's SSD-B and other parameters only available through global challenge authentication. For this reason, the network must support global challenge authentication for voice privacy to be activated.

Voice privacy is enabled as follows. The mobile first calculates the VPMASK based on its SSD-B and global challenge authentication parameters. The VPMASK may be generated at the time of global challenge response so the mobile is prepared to activate voice privacy immediately. The mobile requests voice privacy at the time of call origination or page response on a DCCH or ACC. The Voice Mode information element is included in the origination or page response message to request activation of voice privacy on a DCCH. Similarly, a digital privacy specification is included in Word B of the Origination or Page Response message on an ACC to request activation of voice privacy. If the mobile's SSD is not shared with the serving network, the serving network formulates a TIA/EIA-41 AUTHREQ message and sends it to the mobile's HLR/AC. The AC calculates the VPMASK and returns it to the serving network in the TIA/EIA-41 authreq message. If the SSD is shared, this TIA/EIA-41 exchange is not required. The serving network indicates to the mobile that voice privacy is activated in the Voice Mode information element of the Digital Traffic Channel Designation message on the DCCH or the PM field in the First Digital Channel Assignment Word on the ACC. At this time, voice privacy is activated and the conversation is encrypted.

A mobile may request a change in the voice privacy mode (turning it on or off) after assignment to a DTC. It may also request activation of voice privacy while in conversation on an AVC. To honor this later request, the serving network must handoff the mobile to a DTC. Voice privacy may be continued after an intersystem handoff if both networks support the TIA/EIA-41 signaling to allow the VPMASK to be transferred between them.

Signaling message encryption allows sensitive subscriber information to be encrypted on the DTC and AVC. A limited number of messages that may be sent by the mobile and network while the mobile is in the conversation state may be encrypted. Encryption of these messages prevents an eavesdropper from obtaining information such as the calling party number, in-call feature activation, and any user-entered PIN. The mobile user does not have control over the activation or deactivation of signaling message encryption; it is a service provider option.

A plaintext message and a signaling message encryption key (SMEKEY) are input into an encryption algorithm to generate the cyphertext that is the encrypted message. The SMEKEY is generated in a manner similar to the VPMASK—using global challenge authentication parameters and SSD-B as input to the CAVE algorithm. Signaling message encryption is activated on a per-call basis from either the DCCH or the ACC. The same message flows used to generate the VPMASK and apply voice privacy are also used to generate the SMEKEY and apply signaling message encryption. The mobile is informed that signaling message encryption has been applied to a message through signaling on the DCCH, ACC, DTC, or AVC. Signaling message encryption may be activated at the time of call establishment or during the conversation. It may also be deactivated during a call.

16.6 Encryption on the DCCH

Encryption on the DCCH will be used in the future to protect point-to-point layer 3 messages sent between the mobile and the network. These include most of the layer 3 messages on the SPACH and RACH. Of the most importance is user data such as that contained in teleservices within the R-DATA message. Also important to encrypt is the temporary mobile station identity (TMSI) that may be provide to a mobile within a Registration Accept message. The initial Registration and Authentication messages sent by a mobile on the RACH may not be encrypted during initial system access, because the network may not have the decryption key at that time.

Encryption is supported on the DCCH through signaling at layer 2 on the SPACH and RACH. A 4-bit Extension Header is added to a layer 2 SPACH or RACH frame to indicate that the layer 3 contents of the message are encrypted. The Extension Header identifies the message encryption algorithm (MEA) and message encryption key (MEK) used to encrypt the layer 3 payload. At the receiving end of the layer 2 frame, the layer 3 contents are decrypted based on the Extension Header information. This allows any layer 3 contents to be encrypted, including any new messages developed for a future revision of the Digital PCS standard.

Up to four MEAs and four MEKs may be supported in Digital PCS, with one of the MEAs and MEKs reserved for proprietary implementations. A mobile determines the MEAs and MEKs that are supported on a network by reading the Service Menu message on the F-BCCH or E-BCCH. The mobile may then encrypt layer 3 messages on the RACH using one of the supported MEAs and MEKs.

References

[1] Brown, D., "Techniques for Privacy and Authentication in Personal Communication Systems," *IEEE Personal Communications*, Vol. 2, No. 4, Aug. 1995, pp. 6–10.

[2] Telecommunications Industry Association, TIA/EIA IS-136, *800 MHz TDMA Cellular-Radio Interface—Mobile Station—Base Station Compatibility Standard*, Dec. 1994, Section 6.

[3] Gallagher, M. D., and R. A. Snyder, *Mobile Telecommunications Networking with IS-41*, New York: McGraw-Hill, 1997, pp. 183–223.

Network Parameter Settings

17.1 Importance of network parameter settings

Network parameter settings impact how a mobile accesses and uses services available on a cellular system. These settings can positively or negatively influence a mobile user's perception of the system, the service provider, and the mobile. They have the potential to greatly impact the performance of the mobile in the areas of standby and talk time, selection and reselection time, and quality of service. A high quality of service equates to timely and reliable delivery of information (user data and signaling) to and from the mobile user, a low percentage of blocked calls, good voice quality during a call, and a low percentage of dropped calls.

Network parameter settings are also important to optimize the performance of the network. Less than optimal settings can lead to inefficient operation of the network. For example, some parameter settings may result in too few channels or other resources available for use in one sector while much more than required is available in another sector.

Other parameter settings may result in inefficient use of the forward and reverse DCCH for delivering information to and from mobiles. As a result, the service provider may install more resources than actually needed in order to keep the quality of service high for the mobile user.

Key network parameters for Digital PCS include: (1) the location of DCCHs and mechanisms to assist mobiles in finding DCCHs, (2) access thresholds and retry settings on the DCCH and the DTC that govern when and at what power level a mobile may use one of these channels, (3) the structure of the DCCH, and (4) reselection and handoff boundaries. Table 17.1 lists many of these settings and gives a range of allowable values for each. Network suppliers typically furnish default values for many Digital PCS network parameters to service providers using their equipment, and often provide guidelines for setting the parameters differently based on the situation. For example, mobile access power settings are typically different for picocells than for macrocells. Some parameter settings may even be fixed by the network supplier so that the service provider may not change them to a value that causes poor mobile or network performance. Some service providers may appreciate the network supplier optimizing parameter setting for them, while others would prefer the flexibility to be more creative (that is, give me enough rope to hang myself).

Table 17.1
Key Network Parameters and Range of Values

Parameter	Meaning	Range
Probability blocks	The relative probability associated with finding a DCCH within a block of channels in an 800-MHz band.	16 probability blocks per 800-MHz band
Sub-band priorities	The priority associated with finding a DCCH within a block of channels in a 1,900-MHz band.	7 sub-bands per 1,900 MHz
RSS_ACC_MIN	Minimum received signal strength at the mobile required to access a DCCH.	-111 dBm to -51 dBm in 2-dB steps, and $-\infty$

Network Parameter Settings

Parameter	Meaning	Range
MS_ACC_PWR	Mobile transmit power level required to access a DCCH.	−4 dBm to 36 dBm in 4-dB steps
Access Burst Size	Indication to the mobile whether to use normal or abbreviated burst when accessing a DCCH.	Normal or Abbreviated
Max Retries	Maximum number of access attempts a mobile may make before it considers the access to have failed.	1 to 8 access attempts allowed
Max Busy/Reserved	Indication of the maximum number of times that the mobile can detect the Busy/Reserved/Idle (BRI) field not equal to idle during any given access attempt before declaring an access attempt failure.	1 or 10 BRI ≠ idle
Max Repetitions	Maximum number of times a specific burst within any given access attempt may be sent by the mobile before declaring an access attempt failure.	0, 1, 2, or 3 repetitions allowed
Max Stop Counter	Maximum number of times that either of the following conditions can be detected for any given burst of an access attempt before the mobile declares an access attempt failure: (1) BRI set to reserved or idle after sending an intermediate burst of an access attempt, (2) received/not received (R/N) set to not received along with BRI set to reserved or idle after sending the last burst of an access attempt.	1 or 2 occurrences allowed
Number of F, E, S, reserved, and non-PCH subchannel slots	The number of F-BCCH, E-BCCH, S-BCCH, reserved, and non-PCH subchannel slots per superframe, which determines the number of slots that may be used for addressing mobiles.	No. of F: 3 to 10 slots No. of E: 1 to 8 slots No. of S: 0 to 15 slots No. of Reserved: 0 to 7 slots No. of Non-PCH: none, 2, 4, or 6 slots

Table 17.1 (continued)

Parameter	Meaning	Range
CELLTYPE	An indication of the preference of one DCCH over another control channel, used by the mobile station during the reselection process.	Regular, Preferred, or Nonpreferred
RESEL_OFFSET	The difference in received signal strength required between the current DCCH and a candidate control channel before the mobile may reselect to the candidate.	−128 dB to 126 dB in 2-dB steps
SS_SUFF	The minimum received signal strength required from a preferred candidate control channel before the mobile may reselect to it, or the received signal strength on the current DCCH below which a mobile may reselect to a nonpreferred candidate control channel.	−111 dBm to −51 dBm in 2-dB steps, and −∞
DELAY	On the current DCCH, the length of time the mobile must obtain service before considering reselection to candidate control channels. With respect to a candidate DCCH, the length of time the mobile must wait before considering this candidate for reselection.	0 to 105 superframes (SFs) in 15-SF steps; 105 to 420 SFs in 45 SF steps.
SCANINTERVAL	Basic interval between consecutive signal strength measurements by a mobile on candidate control channels for reselection purposes.	1 to 16 hyperframes, in 1-hyperframe steps
HL_FREQ	Modifier to SCANINTERVAL that indicates to the mobile whether to perform a signal strength measurement on an individual candidate control channel once per superframe (HL_FREQ = 1) or every other superframe (HL_FREQ = 0).	Low or High

17.2 Selection and access parameters

There are a number of network parameters under the control of the service provider that define how a mobile selects and accesses a DCCH. These include the placement of DCCHs, DCCH location information, signal strength and transmit power thresholds for accessing a channel, access burst size and access attempt information, and the structure of the DCCH.

17.2.1 DCCH location

The service provider must decide upon a frequency plan for DCCHs [1]. Unlike ACCs, which are located in a fixed-frequency range in each of the two 800-MHz bands, the Digital PCS standard allows DCCHs to be located anywhere in frequency. This allows for great flexibility in network design, but makes it more difficult for mobiles to locate DCCHs. A probability block structure was standardized to aid mobiles in finding DCCHs from power-up or loss of the radio link. Channels are segmented into 16 probability blocks in each of the 800-MHz bands, as shown in Table 17.2. A mobile scanning for DCCHs in one of the 800-MHz bands typically begins in the highest probability block and scans one block at a time for DCCHs. Therefore, if the service provider can locate most or all of the DCCHs in the highest or second highest probability block, most mobiles will find DCCHs rapidly from power-up.

Table 17.2
800-MHz Probability Blocks

Channel Number	Band	Number of Channels	Relative Probability	Channel Number	Band	Number of Channels	Relative Probability
1–26	A	26	4	334–354	B	21	16 (lowest)
27–52	A	26	5	355–380	B	26	15
53–78	A	26	6	381–406	B	26	14
79–104	A	26	7	407–432	B	26	13
105–130	A	26	8	433–458	B	26	12

Table 17.2 (continued)

Channel Number	Band	Number of Channels	Relative Probability	Channel Number	Band	Number of Channels	Relative Probability
131–156	A	26	9	459–484	B	26	11
157–182	A	26	10	485–510	B	26	10
183–208	A	26	11	511–536	B	26	9
209–234	A	26	12	537–562	B	26	8
235–260	A	26	13	563–588	B	26	7
261–286	A	26	14	589–614	B	26	6
287–312	A	26	15	615–640	B	26	5
313–333	A	21	16 (lowest)	641–666	B	26	4
667–691	A'	25	3	717–741	B'	25	3
692–716	A'	25	2	742–766	B'	25	2
991–1023	A"	33	1 (highest)	767–799	B'	33	1 (highest)

As can be noted from Table 17.2, the highest probability block is located in the expanded spectrum and the lowest probability block is located where the fixed ACCs are assigned. This allows service providers with existing AMPS spectrum to more easily deploy DCCHs in existing frequency plans. The expanded spectrum is traditionally where AMPS service providers begin to deploy digital service.

Sub-bands are defined instead of probability blocks for the 1,900-MHz bands. A probability block is composed of from 21 to 33 channels in the 800-MHz band, corresponding to a frequency range of 630 kHz to 990 kHz for each block. A sub-band is a much larger frequency range, however, as shown in Table 17.3. A sub-band is typically 83 channels, or 2.5 MHz. A mobile begins scanning for DCCHs in the first sub-band in a 1,900-MHz band. Table 17.3 shows the default priority for sub-bands. However, the priority of each sub-band can be programmed into the mobile for intelligent roaming, as described in Chapter 15. A service provider should locate DCCHs in a high-priority sub-band to aid a mobile in rapidly finding a DCCH in the 1,900-MHz band.

Network Parameter Settings 293

Table 17.3
1,900-MHz Sub-Band Priorities

Band	Bandwidth (MHz)	Number of Channels	Boundary Channel Number	Sub-Band Channel Number	Default Sub-Band Priority	Transmitter Center Frequency (MHz) Mobile	Base
Not used		1	1				
A	15	497	2	2–83	4	1,850.010	1,930.050
				84–166	3	1,850.040	1,930.080
				167–250	2		
				251–334	1		
				335–418	5		
				419–498	6		
A and D		1	499		7	1,864.950	1,944.990
A and D		1	500		7	1,864.980	1,945.020
A and D		1	501		7	1,865.010	1,945.050
D	5	164	502	502–583	2	1,865.040	1,945.080
				584–665	1		
D and B		1	666		7	1,869.960	1,950.000
D and B		1	667		7	1,869.990	1,950.030

Table 17.3 (continued)

Band	Bandwidth (MHz)	Number of Channels	Boundary Channel Number	Sub-Band Channel Number	Default Sub-Band Priority	Transmitter Center Frequency (MHz) Mobile	Transmitter Center Frequency (MHz)
B	15	498	668	668–750	4	1,870.020	1,950.060
				751–833	3		
				834–916	2		
				917–999	1		
				1,000–1,082	5		
				1,083–1,165	6		
B and E		1	1,166		7	1,884.960	1,965.000
B and E		1	1,167		7	1,884.990	1,965.030
E	5	165	1,168	1,168–1,250	2	1,885.020	1,965.060
				1,251–1,332	1		
E and F		1	1,333		7	1,889.970	1,970.010
E and F		1	1,334		7	1,890.000	1,970.040
F	5	164	1,335	1,335–1,416	1	1,890.030	1,970.070
				1,417–1,498	2		
F and C		1	1,499		7	1,894.950	1,974.990
F and C		1	1,500		7	1,894.980	1,975.020
F and C		1	1,501		7	1,895.010	1,975.050

Network Parameter Settings 295

Band	Bandwidth (MHz)	Number of Channels	Boundary Channel Number	Sub-Band Channel Number	Default Sub-Band Priority	Transmitter Center Frequency (MHz) Mobile	
C	15	497	1,502	1,502–1,584	6	1,895.040	1,975.080
				1,585–1,667	5		
				1,668–1,750	1		
				1,751–1,833	2		
				1,834–1,916	3		
				1,917–1,998	4		
Not used		1	1,999		7	1,909.950	1,989.990

Besides using brute-force scanning of the 800-MHz probability blocks for DCCHs, another method often used by a mobile while scanning for service in an 800-MHz band is to find an ACC and look for a pointer to a DCCH. Because the ACCs are located in a fixed range of channels, a mobile can quickly and easily scan for the highest signal strength ACC, as described in Chapter 1.

A service provider with Digital PCS and AMPS service can broadcast a message on the ACCs called the Control Channel Information message. This message provides mobiles with the channel number and half the digital verification color code (DVCC) of a DCCH in the same sector as the ACC. The mobile can then move directly from the ACC to the DCCH. The service provider should set the periodicity of the Control Channel Information message on the ACC at approximately once ever five seconds to ensure that all AMPS mobiles can still access the system, as described in Chapter 18. As there are no ACCs defined in the 1,900-MHz bands, it is even more important for the service provider to locate DCCHs in high-priority sub-bands at 1,900 MHz.

Other methods exist for mobiles to find DCCHs upon power-up and call release. The service provider should ensure that these methods are available in the network and are providing the appropriate information to mobiles. The coded DCCH locator (CDL) is present in every $\pi/4$ DQPSK modulated DTC. If a mobile finds a DTC while scanning for a DCCH, it may read the CDL and determine a range of channels (8 channels at 800 MHz or 16 channels at 1,900 MHz) where a DCCH is located. Upon call release from a DTC or AVC, the network may provide a Digital PCS mobile with the exact channel number and DVCC of a DCCH in the same sector as the DTC or AVC. This allows the mobile to rapidly find a DCCH upon call release without entering a probability block or sub-band scan, or scanning the ACCs.

17.2.2 Access thresholds

Access thresholds ensure that adequate signal strength is available for both the mobile and the base station for transactions on a DCCH. The two key access thresholds are RSS_ACC_MIN and MS_ACC_PWR [2]. RSS_ACC_MIN defines the minimum received signal strength at the mobile for a DCCH to be considered suitable for service.

MS_ACC_PWR defines the transmit power that the mobile must use when accessing a DCCH. Both parameters are used by the mobile in the control channel selection process (see Chapter 8) to determine if a DCCH is suitable from a signal strength perspective. When a mobile finds a DCCH in any manner other than reselection, it must guarantee that the candidate meets the following two conditions:

$$RSS - RSS\_ACC\_MIN - \max(MS\_ACC\_PWR - P, 0) > 0 \text{ dBm} \quad (17.1)$$

$$(MS\_ACC\_PWR \leq 4 \text{ dBm AND P\_Class} = 4) \quad (17.2)$$
$$\text{OR } MS\_ACC\_PWR \geq 8 \text{ dBm}$$

where RSS is the received signal strength at the mobile in dBm, and P is the maximum nominal output power of the mobile in dBm as defined by it power class, P_Class.

Equation (17.1) means that the received signal strength on the DCCH must be greater than the minimum received signal strength required to access the cell, plus a bias. The bias is zero if the mobile is capable of transmitting at or above the power level required to access the cell (MS_ACC_PWR), and greater than zero if the mobile is not capable of this output power. In other words, a mobile that cannot transmit at MS_ACC_PWR has to received the DCCH at a stronger signal level than one that can, which means it has to be closer to the cell site. This allows the base station to receive the lower powered mobile with adequate signal strength to decode its messages.

Equation (17.2) means that if MS_ACC_PWR is set to 4 dBm or below, the mobile must be capable of power class IV operation. In other words, the mobile must be capable of transmitting at the required power level for extremely low-powered cell sites, such a picocells. This ensures that mobiles not capable of operating at these low power levels do not access the DCCH and cause interference.

It is easy to see that RSS_ACC_MIN and MS_ACC_PWR set the cell boundaries throughout a Digital PCS system. This makes the proper setting of these two parameters extremely important for the service provider. A service provider with both a Digital PCS and an embedded

AMPS system would typically set these access thresholds so that the coverage of the DCCH matches that of the ACC for every cell or sector. Figure 17.1 illustrates how the combination of RSS_ACC_MIN and MS_ACC_PWR can be used to set the cell size differently. Two examples are shown in the figure, both assuming a power class IV mobile capable of 28-dBm transmit power. The first example shows how the cell radius decreases with higher values of RSS_ACC_MIN if

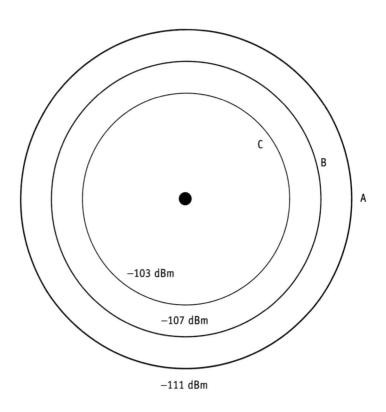

Example 1
MS_ACC_PWR = 28 dBm

A: RSS_ACC_MIN = −111 dBm
B: RSS_ACC_MIN = −107 dBm
C: RSS_ACC_MIN = −103 dBm

Example 2
RSS_ACC_MIN = −111 dBm

A: MS_ACC_PWR <= 28 dBm
B: MS_ACC_PWR = 32 dBm
C: MS_ACC_PWR = 36 dBm

Figure 17.1 Effects of RSS_ACC_MIN and MS_ACC_PWR on cell size for a power class IV mobile.

MS_ACC_PWR is fixed at 28 dBm. The second example shows how the cell radius decreases with higher values of MS_ACC_PWR if RSS_ACC_MIN is fixed at −111 dBm. Except for large cells that expect a high percentage of class I or II mobile traffic, MS_ACC_PWR should typically be set at 28 dBm or lower.

Settings similar to MS_ACC_PWR are provided for the ACC, AVC, and DTC. The control mobile attenuation code (CMAC) parameter specifies the maximum power at which a mobile may access the reverse ACC. On an AVC, the voice mobile attenuation code (VMAC) commands the mobile to transmit at the specified level upon initial channel assignment. The digital mobile attenuation code (DMAC) performs this function on a DTC. No parameter similar to RSS_ACC_MIN exists for these channels, however.

17.2.3 Access attempts

Access attempt information broadcast on the F-BCCH also impacts mobile performance and DCCH capacity [3]. The Access Burst Size defines if a mobile is to use theNormal or Abbreviated length frame format when making an access attempt on the RACH. The Abbreviated length frame format allows for 22 fewer bits of layer 3 data to be transmitted on the RACH per time slot than the Normal length frame format. This could result in the mobile requiring more accesses on the RACH to transmit the same amount of data when using the Abbreviated length frame format instead of Normal length frame format. The more accesses each mobile must make, the fewer RACH slots that are available for other mobiles to use. Also, as the number of accesses required of a mobile increases, the higher the probability that the mobile will have to retransmit one or more of the bursts due to radio link impairments. On the other hand, the abbreviated length frame format is valuable for large-radius cells to avoid collisions on the RACH between mobiles at the cell boundary and mobiles close to the cell site.

Table 17.4 shows the number of bursts for a mobile to transmit typical messages on the RACH using Abbreviated and Normal length frame formats. The mobile is assumed to use a 34-bit mobile identification number (MIN) as its identity in each burst. The Registration message is 21 bits in length with no optional information elements. The Serial

Number message is 21 bits in length. The Page Response message is 32 bits in length with one optional information element to specify the vocoder to use on a call. The Origination message is 89 bits in length with a 10-digit called party number and one optional information element to specify the vocoder. The MACA Report message is 72 bits in length with long-term MACA and short-term MACA for eight channels. The Capability Report message is 65 bits in length with no optional information elements. The combination of layer 3 messages that require the mobile to send more bursts using the Abbreviated length frame format instead of the Normal length frame format are highlighted in bold. The problem is exacerbated for mobile-originated teleservices that use the layer 3 R-DATA message due to the longer length of these messages compared to the ones shown in the table. The moral for the service provider is to set the Access Burst Size to Normal whenever possible.

Table 17.4
Number of RACH Bursts for Typical Layer 3 Messages as a Function of Access Burst Size

Concatenated Layer 3 Messages	Normal Burst Frame Sequence	Abbreviated Burst Frame Sequence
Registration + Serial Number (61 bits)	BEGIN + END (2 bursts)	BEGIN + END (2 bursts)
Page Response + Serial Number (72 bits)	BEGIN + END (2 bursts)	BEGIN + END (2 bursts)
Origination + Serial Number (129 bits)	BEGIN + END (2 bursts)	BEGIN + CONTINUE + END (3 bursts)
Registration + Authentication + Serial Number (101 bits)	BEGIN + END (2 bursts)	BEGIN + END (2 bursts)
Page Response + Authentication + Serial Number (112 bits)	BEGIN + END (2 bursts)	BEGIN + CONTINUE + END (3 bursts)
Origination + Authentication + Serial Number (169 bits)	BEGIN + CONTINUE + END (3 bursts)	BEGIN + CONTINUE + END (3 bursts)

Concatenated Layer 3 Messages	Normal Burst Frame Sequence	Abbreviated Burst Frame Sequence
Registration + Authentication + Serial Number + MACA Report + Capability Report (238 bits)	BEGIN + CONTINUE + END (3 bursts)	BEGIN + CONTINUE + CONTINUE + END (4 bursts)
Page Response + Authentication + Serial Number + MACA Report (184 bits)	BEGIN + CONTINUE + END (3 bursts)	BEGIN + CONTINUE + CONTINUE + END (4 bursts)
Origination + Authentication + Serial Number + MACA Report (241 bits)	BEGIN + CONTINUE + END (3 bursts)	BEGIN + CONTINUE + CONTINUE + END (4 bursts)

Four network parameters govern the number of times a mobile may attempt to access the RACH before giving up on the access attempt: Max Retries, Max Busy/Reserved, Max Repetitions, and Max Stop Counter. These access parameters are broadcast on the F-BCCH, and apply to every mobile that is operating on the DCCH. A distinction must be made between an *access attempt* and a *burst*. An access attempt is defined as the entire transaction that the mobile is attempting to make on the RACH. For example, each of the rows in the first column of Table 17.4 constitutes an access attempt. A burst is one transmission within an access attempt, and fits within one RACH time slot. The rows in the second and third columns of Table 17.4 show the number of bursts required for each access attempt, assuming no retries are required.

The access parameters apply when retries are required on the RACH. Max Retries sets the maximum number of access attempts that a mobile may make on the RACH, and Max Repetitions sets the maximum number of times any burst within an access attempt may be repeated. An access attempt fails if any given burst is not successfully sent after Max Retries attempted retries. For a random access (contention-based) attempt, the mobile cannot receive the BRI field as Busy or Reserved more than Max Busy/Reserved times when attempting to send the first burst of an access attempt before declaring the access attempt a failure. The mobile also cannot detect one of the following conditions more than Max Stop Counter consecutive times before declaring an access attempt failure: (1) the BRI field is set to Reserved or Idle after sending an intermediate burst of an access attempt, or (2) the R/N field is set to Not Received and the BRI field set to Reserved or Idle after sending the last burst of an access attempt.

In most circumstances, it is beneficial to the service provider to set these access parameters to their maximum values. This gives the mobile a fighting chance at completing a transaction on the RACH in a harsh radio environment. A service provider might not use the maximum values for these access parameters if the RF environment is not as harsh (good C/I and SNR over the desired coverage area) and there is concern over RACH capacity. Setting the access parameters lower will not give the mobile as many opportunities to transmit retries of individual bursts or the entire access attempt, resulting in fewer collisions on the RACH.

17.2.4 DCCH structure

The structure of the DCCH to a large extent determines the capacity of the DCCH [4]. It can also impact mobile performance. The structure of the DCCH is defined by the number of time slots per superframe devoted to use as F-BCCH, E-BCCH, S-BCCH, SPACH, and reserved slots. The more slots per superframe that are devoted to the broadcast control channels, the fewer time slots that are available for point-to-point messages such as pages to individual mobile. The service provider must therefore weigh the importance of broadcast information against point-to-point messages when designing the network.

Because information on neighboring control channels is included on the E-BCCH, a lengthy E-BCCH that extends over multiple superframes may cause dragged reselections, particularly in smaller cells. Dragged reselection increases the likelihood that the mobile accesses the wrong DCCH and that a call is either not established or quickly dropped if it is established.

The number of time slots per hour that can be used as PCH slots (NP) is given by the following equation:

$$NP = 5625(32 - NB - NNP) \tag{17.3}$$

where NB is the number of slots devoted to the broadcast channels and reserved slots, and NNP is the number of non-PCH subchannel slots. From Table 17.1, NB is between 4 and 30, and NNP is 0, 2, 4, or 6. A range for the number of PCH slots per hour per DCCH is plotted in Figure 17.2 based on (17.3). This information may be used by the service provider, along with the most appropriate traffic model and retry assumptions, to estimate the capacity of the forward DCCH for various DCCH structures.

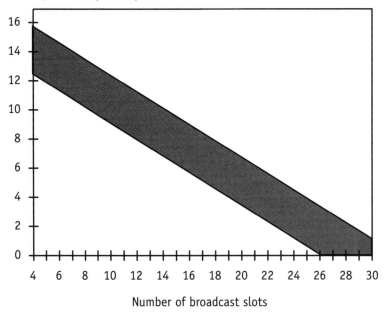

Figure 17.2 Number of PCH slots per hour as a function of DCCH structure parameters.

17.3 Reselection and handoff boundaries

Reselection and the network parameters that impact it are described in detail in Chapter 9, and mobile assisted handoff (MAHO) is described in Chapter 11. This section discusses the alignment of reselection and handoff boundaries. Alignment of these boundaries is beneficial because it ensures that mobiles are operating on the most appropriate sector or cell whenever possible. If a call ends in one sector and the mobile is forced to perform a reselection from the DCCH in that sector to a DCCH in a second sector due to poor signal strength conditions, then the mobile has wasted resources and time. Similarly, and more importantly, if a mobile establishes a call from one sector but is immediately handed off to another sector, the voice quality could be impacted in the process. Another reason to align reselection and handoff boundaries is due to the services that may be offered on one cell or sector, but not offered on a neighboring one. This is particularly true for nonpublic service (see Chapter 14). If a mobile user is expecting one billing rate while operating in nonpublic mode on one cell or sector and another billing rate outside of his or her nonpublic service area, then it is important for the service provider to give the mobile user a clear line of demarcation between these two areas. Of course, reselection and handoff boundaries cannot be set to *exact* locations due to the vagaries of the RF environment. However, a mobile user attempting a call from one location should expect that a high percentage of the time the call will be set up on the same cell or sector and not be immediately handed off to another one.

Aligning reselection and handoff boundaries is tricky for a few reasons. First, the mobile is in control of the reselection process, within the boundaries of the reselection parameters. The mobile makes all signal strength measurements on neighboring control channels and decides the best time to reselect, typically based on signal strength measurements. Handoff, on the other hand, is completely under the control of the network. The network typically uses information from the mobile such as channel quality measurements on the current DTC and neighboring channels to assist it in making the handoff decision. Other variables enter into the handoff decision, however, such as availability of DTCs and trunking facilities in a neighboring sector or cell, the level of interference experienced on the current DTC and the anticipated level on another

DTC, and signal strength measurements made on the mobile's transmissions by the current base station and handoff candidates. All of this makes it difficult to align reselection and handoff boundaries. Add to this the fact that not all networks have the concept of different CELLTYPEs or network types for handoff, but they do for reselection, and the task gets even harder.

To align the reselection and handoff boundaries, the parameters in the handoff algorithm must be set in such a way that the handoff algorithm emulates the reselection algorithm to the greatest extent possible without significantly impacting voice quality. Since handoff algorithms are proprietary to the Digital PCS network supplier, the way to accomplish this differs from one network supplier to another. Some good rules of thumb include setting the MAHO channels to match the neighbor list entries on the DCCH, matching the MAHO and reselection measurement intervals, and matching the DMAC and MS_ACC_PWR settings between DTCs and DCCHs in the same sector or cell.

References

[1] Smith, C., *Practical Cellular and PCS Design*, New York: McGraw-Hill, 1997, pp. 11.52–11.64.

[2] Telecommunications Industry Association, TIA/EIA IS-136, *800 MHz TDMA Cellular-Radio Interface—Mobile Station—Base Station Compatibility Standard*, Dec. 1994, Section 6.

[3] Ibid., Section 6.

[4] Ibid., Section 4.

18

Equipment Testing

18.1 Why test?

Testing is critical to verify network operation and performance. This is especially true for a sophisticated technology such as Digital PCS. If the network is not performing properly, many mobile users could experience poor service. Minor changes to one part of the network can influence operation in the entire network. It is therefore important that any changes to the software or hardware in the network be thoroughly tested prior to commercial release.

Even more critical to test than the network are the mobiles that use it. The mobile is the medium by which the user accesses the network and any services offered by the service provider. If the user experiences a problem with his or her service, the tendency is often to blame the mobile, and the first call is often to the service provider who provided the mobile. It is normally very hard for the service provider to retrieve the mobile if it has a problem. If a bug is found in a particular model of a mobile in the field, then a large effort must often be expended to contact users of that particular model, retrieve their mobiles, and either individually fix them or give the users new mobiles. In contrast, bugs in the network can often be corrected by the service provider in short order. Due to the relative importance of mobile testing over network testing, this chapter concentrates more on the former than the latter.

Finally, testing is needed to verify end-to-end system operation and performance. The system is composed of both the network and the mobiles. These entities may be provided by different manufacturers, and interoperability between them must be achieved for the system to operate properly. If the network manufacturer interprets a particular aspect of the standard one way and the mobile manufacturer interprets it another way, then the two will not be compatible. Interoperability between different networks must also be achieved, so that a mobile user in one network can communicate with a mobile user in another.

18.2 Network testing

Network testing is carried out by both the network supplier and the service provider in the field. It includes verifying that new features or functionalities work properly and that the existing operation is not impacted by any upgrades put in place to provide the new features or functionalities. Equipment used in network testing includes mobiles, cellular system analyzers, network analyzers, and measurement and analysis functionality inherent in the network.

Network testing is required anytime a change to the network occurs. This may include the addition of a new hardware element such as a new radio, or an upgrade to an existing hardware element, such as the addition of more memory to a processing unit. The change may be new software in one or more of the network elements, such as the base station or the mobile switching center (MSC).

For Digital PCS, network testing is often tied to the package release schedule of the network supplier. New features and functionality in the network are typically bundled together and offered to the service provider as a package by the network supplier every three to six months. A few months prior to commercial launch, the network supplier thoroughly tests the new package in the field with a service provider in what is called a first office application (FOA). When the FOA is successfully completed and any bugs in the new package release have been corrected, the new package is commercially available and has met general availability (GA). Any bugs that the service provider finds after GA are typically reported to the network supplier and fixed in the next package release, or sooner if the bug impacts service to a large population of users.

The key network tests to conduct with a new package release are feature testing and regression testing. Feature testing verifies that the feature performs as expected in the network. Feature testing must verify that the service provider can activate the feature, the mobile user can access the feature, the service provider can bill for the feature, and the service provider can gather statistics on the use of the feature. Regression testing verifies that the feature does not adversely impact the overall operation of the network and does not break another feature.

Prior to conducting network testing, test objectives are identified, test cases are developed, and a test plan is written. A test case describes the procedures for conducting a test and the expected outcome. The test plan is composed of the test cases and a description of the equipment and personnel needed for the testing. If the network testing is conducted on a live network, it is typically performed at night. This reduces the likelihood that service will be disrupted while the testing is performed. Figure 18.1 summarizes the network testing process.

18.3 Mobile testing

Mobile testing is carried out by both the mobile supplier and the service provider in the laboratory and in the field. It includes performance testing, interoperability testing, and protocol conformance testing. Service providers typically require mobiles to pass all of these tests prior to commercial availability in their networks. Equipment used in mobile testing includes mobiles, cell site simulators, cellular system analyzers, the cellular network, and measurement and analysis functionality inherent in the network.

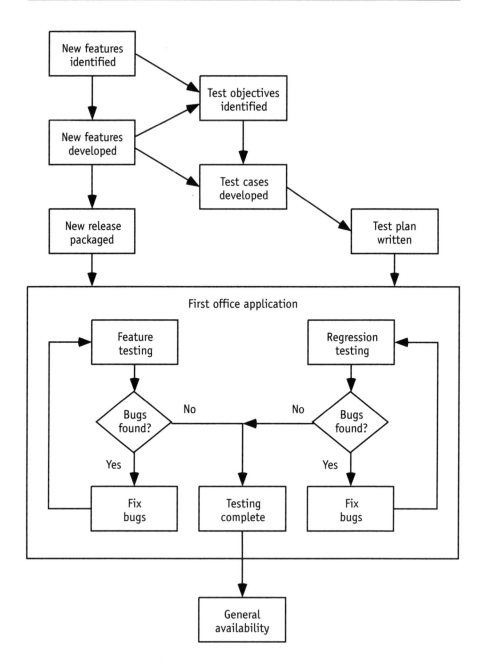

Figure 18.1 Summary of the network testing process.

18.3.1 Performance testing

Performance testing verifies that the physical and link layers of the mobile operate as required by the IS-137 standard. It also verifies that other aspects of mobile performance, such as talk time and standby time, meet both the claims of the mobile supplier and the requirements of the service provider. Key mobile performance parameters that are tested include power output; receiver sensitivity; resistance to interference, Doppler spread, and multipath fading; selection time; reselection time; accuracy of channel quality measurements; talk time; standby time; and audio quality in clear channels, in degraded channels, with background noise, and during a handoff. Except for selection time, reselection time, talk time, and standby time, testing methods for key mobile performance parameters are specified in IS-137. Therefore, the remainder of this section describes methods to test the key mobile performance parameters not covered by IS-137.

Selection time is the length of time from mobile power-up to registration on a control channel. A short selection time is important to keep the mobile user from becoming impatient with the operation of the mobile and the network. When a user powers on his or her phone, he or she typically wants to make a call right away. For AMPS mobiles, selection time is not a critical issue. The ACCs are located in a fixed range of channels in each 800-MHz cellular band. It is a simple matter for an AMPS mobile to quickly scan the 21 ACCs in a cellular band, select the strongest one, read it, and register if appropriate. This process is described in greater detail in Chapter 1. Selection to a control channel is a more complex process for a Digital PCS mobile for two reasons. First, a DCCH may be located on any channel in a particular band. Where does the mobile begin to search for a DCCH? Second, Digital PCS allows for mobiles to be capable of dual-band 800-MHz and 1,900-MHz hyperband operation. What band order should a mobile use in scanning for a DCCH?

The mobile supplier has a number of tricks at his or her disposal to minimize the selection time. This includes keeping a history of the last-used DCCHs and scanning them first, making rapid signal strength measurements and synchronization attempts, and quickly identifying DCCHs of acceptable service providers.

Measuring selection time is a straightforward process of powering on a mobile and measuring the time until the mobile receives a registration accept message on a DCCH and displays an in-service indication. Many mobiles have a special test mode from which the current state of the mobile can be quickly determined. Selection time measurements should be repeated a number of times to get a good statistical sample of data, and should be averaged over the entire sample. The hard part in selection time measurement is setting up the scenarios under which the tests are conducted. The best way to conduct selection time testing is in the field where multiple DCCHs are available, and the RF environment can be changed by changing locations. This type of testing in the field also takes the longest time to complete, however. Laboratory testing of selection time can be completed rapidly if adequate cell site simulation capability is available.

Reselection time is the length of time the mobile takes to change from operation on a DCCH to operation on a neighboring control channel, either a DCCH or an ACC. Reselection time is critical because if a mobile is not on the optimal control channel at any given time, a call may be set up from a control channel with poor RF characteristics. The user may experience poor call quality or even a dropped call as a result. Reselection time is governed by a number of parameters broadcast on the current DCCH, including various delays and biases described in detail in Chapter 9. The key to reselection testing is to remove the reselection delays and biases as variables so that only the performance of the mobile is evaluated. The accuracy and speed of the signal strength measurements on the current and neighboring control channels differentiates the reselection performance of one mobile from another.

The most repeatable way to perform reselection testing is in the laboratory using two cell site simulators and a mobile cabled into both through combiner/splitter circuitry. This allows for accurate control of the RF levels of the two control channels being transmitted by the cell site simulators. The RF levels of the two control channels are changed to make the mobile reselect from the DCCH to the neighboring control channel, and the time from the RF level change to the instant when the mobile settles on the new control channel is measured. It is once again useful if the mobile can be placed in test mode to observe exactly when the mobile reselects from one channel to the other. The test should be repeated a number of times and the results averaged to yield a good approximation of the selection time.

Talk time is the total time that a mobile can be in the Conversation state before its battery is drained and the mobile loses power. Talk time is an important marketing feature of a mobile. The longer the talk time, the less the user must worry about changing or charging batteries. Talk time for Digital PCS mobiles is quoted separately for operation in analog mode and in digital mode. It can vary drastically based on the power level at which the mobile is operating and the size of the battery used with the mobile.

Talk time tests should be conducted with a fully charged battery—typically the standard battery that comes with the mobile. The mobile power level on the DTC or AVC should be set to the most typical value experienced in the system. A good setting to use is 28 dBm, the maximum output power of the mobile. This gives a worst-case talk time. The easiest way to perform a talk time test is in the laboratory with a cell-site simulator. Most mobiles contain a call timer that logs the time of the last calls. A call is initiated between the mobile and the cell site simulator, and allowed to continue until the mobile powers down. After the test, a fresh battery is placed on the mobile and the call timer is queried to determine the talk time. Once again, this test should be repeated a number of times to get an accurate reflection of talk time.

Standby time is the total time that a mobile can be in the Idle task or DCCH Camping state on a control channel before its battery is drained and the mobile loses power. Standby time is important to the user for the same reasons as talk time. Standby time for Digital PCS mobiles is also quoted separately for analog mode and digital mode.

A good way to conduct a standby time test is to power-up a mobile with a fully charged battery and log the power-up registration and all periodic registrations until the mobile fails to register periodically. The periodic registration interval can be set from the network of the cell site simulator. A good value to use for the periodic registration interval for standby testing is 10 minutes. This allows standby time to be measured with an accuracy of approximately 10 minutes, which is close enough when standby time is measured in multiple days. Because of the length of time that standby testing takes, it is beneficial to test multiple mobiles at the same time and average the results across all of them.

18.3.2 Interoperability testing

Interoperability testing verifies that products from different suppliers operate correctly together. It is typically carried out in the network suppliers' laboratories. Immediately after the standardization of IS-136, the Digital PCS industry developed an interoperability test plan, *IS-136 Interoperability Test Specification,* available to members of the Universal Wireless Communications Consortium (UWCC), with the objectives of ensuring base station and mobile compatibility and debugging preproduction IS-136 (now TIA/EIA-136) equipment. The scope of the interoperability test plan is limited to basic protocol testing in a laboratory setting. Table 18.1 summarizes the coverage of the interoperability test plan.

As shown in Table 18.1, the interoperability tests are segmented into phases and stages. This allows mobiles and base stations at various stages of development to be tested together. For example, many of the tests in phases 1 and 2 require access to layer 1 and layer 2 operation that is only available in developmental units. The tests in phase 3 are often the only ones appropriate for equipment nearing production quality.

The procedure followed in interoperability testing is for a mobile supplier to schedule testing with a network supplier. The mobile supplier may elect to skip phase 1 and 2 testing if he or she is confident of the lower layer implementation. Testing is conducted by the mobile supplier with assistance from the network supplier. Test results may be reported to the service provider upon request. Successfully passing interoperability testing does not guarantee that the mobile meets performance requirements, or that the mobile will be compatible with future releases of TIA/EIA-136. It does, however, complement other aspects of mobile testing.

In addition to the interoperability testing described above, a service provider may conduct interoperability testing between networks supplied by different network vendors. This type of testing is typically conducted to verify that TIA/EIA-41 has been implemented the same by different suppliers. Interoperability at this level is required to allow mobile users to roam between networks while still obtaining service and access to advanced features such as short message service. It is also

Table 18.1
Summary of Key Mobile Interoperability Tests

Phase	Stage	Summary of Test Cases
Phase 1	Stage 1–Layer 1 slot structure	Verify forward and reverse DCCH slot structures.
	Stage 2–Layer 2 protocols	Verify normal and abbreviated single- and multislot bursts, in ARQ and non-ARQ mode.
	Stage 3–Idle mode	Verify DCCH selection based on a scan of the probability blocks and from reading the DCCH Locator field on the DTC.
Phase 2	Stage 4–Call processing	Verify the mobile is able to originate and receive calls from a DCCH to an AVC and DTC, and that the mobile properly releases from the AVC and DTC.
Phase 3A	Stage 5–Registration and reselection	Verify the mobile appropriately registers to a DCCH on power-up, power-down, and change in location area. Verify that the mobile selects to a DCCH according to information in an AVC and DTC release message, and according to the control channel information message on the ACC. Verify the mobile properly reselects to a new DCCH upon periodic evaluation, radio link failure, cell barring, and directed retry.
	Stage 6–Short message service	Verify various length SMS messages can be delivered to a mobile and that the mobile responds appropriately to SMS setting.
	Stage 7–Advanced features	Verify that authentication, voice privacy, and signal message encryption operate correctly in a mobile, and verify the optional features of calling party number, message waiting indicator, distinctive ringing, and capability request operate correctly.

Table 18.1 (continued)

Phase	Stage	Summary of Test Cases
	Stage 8–Functional/protocol evaluation, call scenarios	Verify mobile-to-land, land-to-mobile, and mobile-to-mobile call setup with different combinations of voice services. Verify digital-to-analog, digital-to-digital, analog-to-digital, and analog-to-analog handoffs.
Phase 3B	Stage 9–Mandatory/minimum performance	Verify a mobile properly interprets reselection parameters, properly registers with a system, properly sends a MACA report, and transmits at the proper power level.
	Stage 10–ID structure (private and residential systems)	Verify a mobile properly registers with the public network and with a PSID or RSID matching one stored in the mobile.
	Stage 11–Optional functionality	Verify that any optional mobile functionality implemented by a mobile operates correctly, including reduction in scanning frequency of neighbor cells, mobile-originated point-to-point SMS, and paging frame classes greater than 1.
	Stage 12–Revision A functionality	Verify proper operation of one-button callback, priority access and channel assignment, and over-the-air activation.

important to the service provider that billing and authentication are performed correctly when mobile users roam.

18.3.3 Protocol conformance testing

Protocol conformance testing is a method of verifying that a mobile has implemented layer 3 of the Digital PCS standard correctly. Because the proper operation of layer 3 requires that the lower layers of the standard are implemented correctly, protocol conformance testing also verifies layer 1 and layer 2 functionality to a limited extent. Protocol conformance testing is carried out in a laboratory environment using cell site simulators. Cell site simulators can be easily programmed to test aspects of layer 3 that have yet to be implemented in the network. This makes protocol conformance testing a method of ensuring forward compatibility of mobiles when new layer 3 functionality is added to the network. Protocol conformance testing therefore complements interoperability testing.

Protocol conformance testing is required due to the sophistication of the Digital PCS standard in relationship to earlier cellular standards. However, an argument can be made that even mobiles designed to the AMPS standard should undergo protocol conformance testing. If a mobile supplier has misinterpreted the standard and protocol conformance testing has not been conducted on the mobile, then it could be years before the wrong interpretation manifests itself in the operation of the mobile. By that time, many thousands of the noncompliant mobiles may be in use, and it may be impractical for a service provider to implement a new feature in the desired way. A real-life example may explain this point better.

A new message was developed for inclusion in the overhead message train of the ACC to support Digital PCS. The new message, called the Control Channel Information message, identifies the channel number of a DCCH in the same sector as the ACC. The message provides other information to assist a Digital PCS mobile in obtaining service on the most appropriate control channel, including half the DVCC of the DCCH and different features that are available on the DCCH. Reading the Control Channel Information message from an ACC is often the fastest way for a Digital PCS mobile to find a DCCH from power-up. This is because the ACCs are always located on known channel numbers, but the DCCHs are not. The TIA/EIA-136 standard allows the Control Channel Information message to be broadcast on the ACC at an interval from approximately once per second to once every five seconds. The more frequently the message is sent on the ACC, the faster a Digital PCS mobile can read the message and move to the DCCH at power-up.

When Digital PCS field testing began, service providers learned that service would be disrupted for a large number of AMPS mobiles if the Control Channel Information message was broadcast continuously on the ACC. The affected AMPS mobiles would not recognize the Control Channel Information message as being a part of the overhead message train on the ACC, and would not appropriately decrement a counter, called the number of additional words coming (NAWC), which identifies the end of the overhead message train and the beginning of mobile station control messages and control filler words. As a result, the mobiles would go in and out of service on the ACC and had the potential to miss pages. A significant amount of testing was conducted to determine an appropriate periodicity at which the Control Channel Information message could be broadcast on the ACC to help Digital PCS mobiles find a DCCH while ensuring that these AMPS mobiles would remain in service. This testing, and the suboptimal periodicity of broadcasting the Control Channel Information message, could have been avoided had protocol conformance testing been conducted on AMPS mobiles years before.

The Digital PCS industry has developed nine protocol conformance test scenarios, each composed of a multiplicity of individual test cases. A pass/fail criterion is included for each test case. The test cases are written in such a way that the mobile can complete one case and move directly into the next case in the same test scenario. The test scenarios have been combined into a document, the *Consolidated Protocol Test Specification*, available to UWCC members. Table 18.2 summarizes the coverage of the *Consolidated Protocol Conformance Test Plan*. Acronyms are described in the Glossary.

Table 18.2
Mobile Protocol Conformance Test Plan Coverage

Test Scenario	Test Case	Description
1		Basic call setup and release
	1	Mobile station camps on unknown DCCH
	2	Registration on DCCH
	3	Mobile station origination on DCCH assigned to AVC
	4	Audit procedure on AVC
	5	Mobile station release on AVC with DCCH info
	6	Mobile station audit on DCCH
2		AVC and DTC assignments and handoffs

Equipment Testing 319

Test Scenario	Test Case	Description
	1	Mobile station camping on last known DCCH
	2	Registration on DCCH
	3	Mobile station origination on DCCH assigned to AVC
	4	Audit procedure on AVC
	5	Handoff from AVC to DTC
	6	Mobile station release on DTC
	7	SSD update on DCCH
	8	BMI origination on DCCH with authentication
	9	Capability update on DTC
	10	SMS on the DTC
	11	BMI release on the DTC/selection time
	12	Registration accept display on DCCH
	13	Unique challenge on DCCH
	14	BMI originated SMS on DCCH
	15	User alert on DCCH
	16	Parameter update on DCCH
	17	BMI origination on DCCH with authentication assigned to AVC
	18	Mobile station power-down during analog call
3		MACA, TMSI, and second DCCH
	1	Registration on DCCH and TMSI assignment
	2	Verification of mobile station paging frame class
	3	Mobile station origination on DCCH assigned to AVC
	4	Audit procedure on AVC
	5	Handoff from AVC to DTC
	6	Mobile station release on DTC
	7	Broadcast change notification and MACA report
	8	Forced registration on DCCH with TMSI assignment
	9	Mobile station origination with reorder
	10	BMI Page with TMSI and intercept tone
	11	Current cell barred reselection test
	12	Power-down registration
4		Broadcast change notification and reselection
	1	Registration on DCCH with reject
	2	Emergency information broadcast with BCN change
	3	Directed retry on DCCH with location area registration
	4	Reselection based on periodic scanning

Table 18.2 (continued)

Test Scenario	Test Case	Description
	5	Power-down on DTC with automatic release
5		Selection, reselection, and nonpublic mode operation
	1	Mobile station selection performance on power-up, only ACC available
	2	Proper mobile functionality when DCCH pointer is placed on the ACC but no DCCH exists
	3	Reselection to DCCH based on DCCH info on ACC
	4	ACC to DCCH registration
	5	Display of alphanumeric SID
	6	Reselection on periodic scanning
	7	New system registration on private system
	8	PSID/RSID alphanumeric name display
	9	Conversation on DTC
	10	Intersystem dedicated DTC handoff
	11	Selection performance on call release
	12	New system registration and default alpha tag display
	13	Reselection on server degradation
	14	Automatic test registration to acquire PSID/RSID alphanumeric name
	15	Roam display inhibit
	16	Manual system reselection
6		Mobile assisted handoff (MAHO) tests
	1	Mobile station camping and registration on DCCH
	2	BMI origination of DCCH assigned to DTC, DTX disabled
	3	24-channel MAHO configuration on DTC
	4	Stop MAHO operations on DTC
	5	MAHO operation during and after handoff
	6	Handoff back to an IS-136-capable DTC
	7	Verification of unused channel response
	8	Message interruption
	9	Hyperband measurement order
7		Point-to-point teleservices
	1	Mobile station camping on unknown DCCH
	2	BMI originated SMS on the DCCH
	3	BMI originated SMS on the DTC

Test Scenario	Test Case	Description
8		Bad data
	1	Mobile station power-up and registration on a DCCH
	2	Emergency information broadcast via BCN change
	3	Page response on DCCH
	4	Reselection to new DCCH on periodic scanning
	5	Location area registration
	6	Authentication
	7	Mobile station origination on DCCH with MACA report and authentication
9		SPACH interruption and ARQ mode operation
	1	Mobile station power-up and registration
	2	Non-ARQ mode SMS with no interruption
	3	Non-ARQ mode SMS with PCH interruption
	4	ARQ mode SMS with no interruption
	5	ARQ mode SMS with interruption
	6	ARQ mode SMS with ARQ mode SMS interruption
	7	ARQ mode SMS with PCH interruption

Protocol conformance testing is conducted in a laboratory setting. Many of the tests require two cell site simulators. The mobile under test is typically cabled into the cell site simulator(s), which are controlled from a PC. A graphical user interface allows the tester to run the tests, vary the test conditions, and log the results. For example, the user can set the frequency band upon which the tests are to be conducted. A number of the parameters in the tests are controlled by a random seed. This adds variability to the tests and allows for different parameter settings to be used. The random seed used in a test is written to the log file, and the test can be rerun with the same seed, if desired. The test results are written to a test log for postprocessing to determine if the mobile passes or fails.

It should be noted that protocol conformance testing cannot guarantee 100% compliance to the standard unless exhaustive testing is performed. Exhaustive testing would exercise every possible combination of variables that the mobile could encounter. This would be impractical because of the time to develop and run the tests. The product would be obsolete before exhaustive testing was completed! The best that protocol conformance testing can hope to achieve is to catch the most probable bugs before commercial product is released.

18.3.4 CTIA certification program

The Cellular Telecommunications Industry Association (CTIA) certification program tests mobiles in the areas of electrical parameters, ESN security and authentication, and health and safety features. The purpose of the program is to enhance mobile conformance with established industry standards, thereby maximizing user satisfaction with service provider networks. When the CTIA determines that a mobile meets or exceeds the requirements of the industry standards governing the performance of the mobile, the CTIA certification seal is awarded. A service provider not having test capability of his or her own may desire that a mobile obtain the CTIA certification seal prior to commercially releasing the mobile in his or her markets.

The bulk of the CTIA testing is in the area of electrical parameter evaluation. For Digital PCS mobiles, electrical parameter testing is conducted to the requirements of IS-137, as described in Section 18.3.1 of this chapter. The ESN security testing verifies that the ESN cannot be altered outside of a manufacturer's own facilities. The authentication testing is a test of the authentication protocol and verifies that the mobile properly stores, calculates, and uses authentication information. The health and safety features require all mobiles marketed for use in an automobile to operate in a "hands-free" mode, and that complete TIA-approved health/safety language be included in the literature provided with all mobiles. In addition, the CTIA requires mobile manufacturers to certify compliance with the FCC guidelines for RF exposure.

To be considered for award of the CTIA certification seal, two production or preproduction mobiles must be submitted to the CTIA by the mobile supplier. A fee is assessed to the mobile supplier by the CTIA for all certification testing. The CTIA contracts with an external laboratory to perform the required testing according to a test plan developed by the CTIA. Laboratory testing is performed in batches on a monthly basis and takes approximately six to eight weeks to complete. Laboratory test results are provided to the mobile supplier. If the mobile passes testing, the mobile supplier is allowed to label the product with the CTIA certification seal. CTIA certification is granted for a period of 18 months, after which time a mobile must be recertified. The certification can apply to multiple models of a mobile if the mobile supplier can show that the models do not significantly differ in electrical and mechanical characteristics. The CTIA may conduct random testing during the certification period of a mobile to ensure continued conformance to the industry standard.

Towards Global TDMA

A CRITICAL MASS of subscribers must be built for a cellular technology to become cost-effective for manufacturers and service providers. This fact was realized by the proponents of Digital PCS, and activities were initiated in 1995 to make Digital PCS a global standard. An industry organization was formed to ensure that Digital PCS does not remain a North American–only standard. This chapter describes the organization that was formed by the Digital PCS industry to meet this need.

19.1 The Universal Wireless Communications Consortium

The Universal Wireless Communications Consortium (UWCC) is a wireless industry organization consisting of leading Digital PCS vendors and service providers. Appendix A contains a list of UWCC members. The objectives of the UWCC are to (1) promote deployment of Digital

PCS products and services worldwide, (2) stimulate Digital PCS subscriber and service provider growth on a global scale, (3) provide information on Digital PCS to facilitate conversion of AMPS-based systems to Digital PCS, (4) expedite the deployment of Digital PCS into all global markets, and (5) continue the development of enhanced Digital PCS services.

The UWCC conducts a number of activities in support of its business objectives. The UWCC offers regional conferences and global summits around the world to facilitate the exchange of views on technological, commercial, and legislative aspects of Digital PCS. UWCC publications elaborate on the Digital PCS family of standards and provide a tutorial on the technology for new UWCC members. The marketing arm of the UWCC promotes Digital PCS in wireless industry publications. Finally, three UWCC forums oversee the technical aspects of Digital PCS: the Global TDMA Forum, Global Wireless Intelligent Network (WIN) Forum, and Global Operators' Forum. These forums work with the TIA to standardize new Digital PCS services, develop procedures to test these new services, and identify performance metrics for Digital PCS. Membership dues fund the activities of the UWCC.

The UWCC is a limited liability corporation with headquarters in Redmond, Washington. Figure 19.1 shows the organization of the UWCC. A board of directors composed of representatives from the most prominent member companies (board-level companies) sets the policies and direction of the UWCC. An executive director and his or her support staff manages the day-to-day activities of the UWCC. A manager who reports to the executive director heads each UWCC forum. Participation in the forums by UWCC member companies is completely voluntary.

19.2 Global TDMA Forum

The Global TDMA Forum (GTF) focuses on the evolution of the TIA/EIA-136 air interface for Digital PCS. GTF activities include development of requirements for new Digital PCS features and services, development of implementation guides for Digital PCS, interoperability testing and coordination, and representation of the UWCC within TIA (see Chapter 3 for a description of the TIA). Another key activity of the

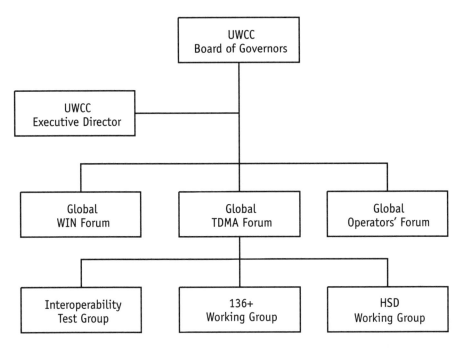

Figure 19.1 UWCC organization chart.

GTF is to provide international manufacturers with an opportunity to influence the standardization of Digital PCS features and services. These activities are carried out within the GTF proper and within three working groups under the GTF. Examples of GTF working groups include the Interoperability Test Group (ITG), the 136+ Working Group, and the High Speed Data (HSD)Working Group. Working groups of the GTF may be created and disbanded based upon specific needs at a given time. A chair reporting to the GTF manager leads each working group.

The ITG develops test specifications for new Digital PCS features and services. The test specifications developed by the ITG supplement the minimum performance requirements and tests defined in IS-137 and IS-138 (see Chapter 3 for a description of these standards). They are used to verify the proper operation of new features and services, and to guarantee the interoperability of mobiles and base stations implementing these new features and services. The test specifications developed by the ITG may be used by service providers and vendors in the laboratory and in the field. The ITG also works with the Cellular Telecommunications Industry Association (CTIA) and with manufacturers to ensure that Digital PCS certification and interoperability testing are carried out to the level of quality required by the UWCC.

The 136+ Working Group and the HSD Working Group have similar roles. The 136+ Working Group was chartered with developing enhanced voice and data services for the TIA/EIA-136 30-kHz channel prior to TIA standardization activities. The 136+ Working Group developed the stage 1 feature description and conducted the technical analysis of alternatives that resulted in the new modulation and vocoder for TIA/EIA-136 described in Chapter 20. The HSD Working Group was chartered with developing a high data rate component of TIA/EIA-136 to meet the requirements of IMT-2000, as defined in Chapter 20. The HSD Working Group conducted the technical analysis of alternative technologies and built consensus around the high-speed data component of TIA/EIA-136 defined in the next chapter.

Another role of the GTF is to develop implementation guides that supplement the Digital PCS family of standards. Implementation guides assist manufacturers and service providers to better understand Digital PCS features and services. They are necessary because important aspects of a feature or service may not be subject to standardization. An example is a feature that relies on the user interface and memory management function of a Digital PCS mobile, such as nonpublic service (described in Chapter 14). An implementation guide has been formulated to assist mobile manufacturers with developing a mobile to meet the nonpublic service requirements of the Digital PCS service providers. The same implementation guide gives service providers additional descriptive information of nonpublic service that is not included in the Digital PCS standards. Other implementation guides that have been formulated by the GTF and that are available to UWCC members include descriptions of short message service, over-the-air activation, data services, and minimum feature and functionality requirements for TIA/EIA-136 mobiles.

19.3 Global WIN Forum

The Global WIN Forum (GWF) focuses on developing new WIN features for Digital PCS based on the TIA/EIA-41 intersystem standard.

Additions to the TIA/EIA-41 standard, or changes to the network architecture, to support new Digital PCS features and services on the TIA/EIA-136 air interface are also the purview of the GWF. WIN defines the ability for service providers to implement custom features without the need for individual feature standardization, including voice-controlled services, incoming call screening, and calling name identification presentation. Some of these features require corresponding TIA/EIA-136 air interface support, such as calling name presentation, while others do not.

19.4 Global Operators' Forum

The Global Operators' Forum (GOF) is responsible for the development of common service provider requirements and operational standards and procedures. The GOF provides input to both the GTF and GWF for activities the service providers desire these forums to pursue. This process helps to ensure that the development of Digital PCS standards includes necessary operational requirements and meets network performance criteria. The GOF also serves as a stage for testing and validating Digital PCS prestandards proposals and recommendations coming from the GTF and GWF. For example, the GOF may conduct early testing of 136+ and HSD capabilities defined in the GTF. The GOF activities result in the development of reports, standards, common practices, and other publications related to Digital PCS operation.

The GOF is structured to support a tactical arm and a strategic arm. The tactical activities of the GOF include the testing and operational aspects of Digital PCS. GOF strategic activities include identifying service provider requirements for new Digital PCS features and services and feeding them into the GTF and the GWF. This assures that consensus is built by the service provider on new enhancements to TIA/EIA-136. It also assures that international service providers that do not have a voice in the TIA standards process are able to influence new Digital PCS feature and service development.

The Future of Digital PCS

20.1 IMT-2000 and third-generation cellular

One must view the wireless landscape from a global perspective when considering the future of Digital PCS. The entire world is caught up in the wireless telecommunication revolution. Economies of scale and the capability for global roaming will be keys to the success of any wireless technology in the future. Enter IMT-2000, which is a third-generation digital cellular system as standardized by the International Telecommunications Union (ITU). IMT-2000 was formerly called the Future Public Land Mobile Telecommunications System (FPLMTS), often pronounced, "flumpts." Perhaps because flumpts did not sound appropriate for a third-generation cellular system, the name was changed to International Mobile Telecommunications-2000 (IMT-2000). The 2000 in IMT-2000 refers to 2,000 MHz, the frequency band where IMT-2000 may operate in much of the world. The third-generation cellular standardization effort by the ITU did not attract true global attention until 1997, when a few major telecommunications equipment suppliers around the world announced they would support a particular technology for IMT-2000, and the prospect for a global cellular standard began to dawn on the world.

Besides the potential for IMT-2000 to be a global system with realizable economies of scale for equipment, the other key attractions of IMT-2000 are its objectives to support high-quality, low-delay voice services and high-speed data services. The primary voice services objectives are to provide performance equivalent to 32-Kbps adaptive differential pulse-code modulation (ADPCM) operated in the fixed network [1], with one-way end-to-end delay of less than 40 ms [2]. The primary circuit and packet data services objectives are to provide user data rates up to 2,048 Kbps in an indoor office environment, up to 384 Kbps in a pedestrian environment, and up to 144 Kbps in a vehicular environment, all at 10^{-6} bit error rate [3].

As envisioned by the ITU, IMT-2000 requires spectrum allocations set aside by the World Administrative Radio Conference (WARC) in the 1,885–2,025-MHz and 2,110–2,200-MHz frequency bands [4]. The 1,850–1,990-MHz frequency band is already allocated for personal communications service in North America, and Digital PCS systems already operate in this spectrum. There is no reason why many of the key services envisioned for IMT-2000 cannot also be provided by enhanced Digital PCS systems operating in the 800-MHz and 1,900-MHz frequency bands. In 1997, the UWCC (see Chapter 19) realized this and began defining enhancements to TIA/EIA-136 to meet the objectives of IMT-2000.

The UWCC defined two sets of enhancements to TIA/EIA-136 to meet the IMT-2000 voice and data services objectives. The first was the addition of a higher level modulation format and a correspondingly higher bit rate vocoder for use in the regular Digital PCS 30-kHz TDMA radio channel to meet the IMT-2000 voice-quality objectives. This set of enhancements is known as 136+. The second set of enhancements was the definition of a 200-kHz TDMA radio channel to meet the IMT-2000 pedestrian and vehicular data rate objectives. This set of enhancements is know as 136++, or 136HS (HS for high speed). While additional functionality was identified for 136+ and the 136HS, the aforementioned enhancements constitute the most important items to make Digital PCS align with IMT-2000 objectives. TIA/EIA-136, 136+, and 136HS together form what is known as UWC-136, which is a third-generation cellular system offering.

20.2 136+

The 136+ enhancement defines two new voice services and a packet data service for Digital PCS using 30-kHz channel bandwidths. One of the voice services uses the IS-641-A ACELP vocoder and DQPSK modulation with more channel coding on the downlink to gain approximately 2 dB of robustness to interference on this link. The greater robustness is achieved by stripping out the SACCH, CDL, and CDVCC fields from the DQPSK time-slot format for the forward DTC and using them for DATA (see Chapter 4 for a description of the fields in the DTC time slot). The other voice service uses a similar slot format with 8-PSK coherent modulation to gain another bit per symbol and allow the inclusion of a 12.2-Kbps ACELP vocoder with even higher fidelity than the IS-641-A vocoder. The 12.2-Kbps ACELP vocoder has 5 ms less processing delay than the IS-641-A vocoder. A one-slot interleaving scheme is also defined to reduce the end-to-end delay by another 20 ms to meet the IMT-2000 delay objective. The forward and reverse time-slot formats for the 8-PSK DTCs include reference symbols that facilitate demodulation at the receiver, since the modulation is coherent and not differential.

The 136+ packet data service may operate with either DQPSK or 8-PSK modulation, depending upon the application. DQPSK modulation is used where a more robust, lower throughput data service is required, while 8-PSK modulation is used where a less robust but higher throughput data service is required. The capability exists to switch between the modulations based on the environment and needs of the mobile user at any given time. In triple-rate mode, a 136+ packet data user is capable of achieving approximately 43.2 Kbps of usable data rate with the 8-PSK modulation format.

A packet control channel (PCCH) and a packet traffic channel (PTCH) are defined in 136+ to carry packet data traffic. The PCCH is similar to the DCCH in that a mobile capable of 136+ packet data service may read the PCCH and obtain system overhead information necessary for registering and authenticating with the packet data network and reselecting to neighboring PCCHs. The PCCH may carry more overhead than is strictly necessary to allow for the possibility of devices that are only capable of packet data service—that is, that do not use the DCCH for signaling and control. The PCCH logical channel structure is similar to the DCCH to allow a mobile to enter sleep mode while operating on a PCCH. The PTCH is similar to the DTC in that a mobile may be assigned to a PTC to carry out a lengthy packet transaction.

A close integration of voice and data services is achieved with 136+ by allowing a mobile to be paged for a packet transaction while in service on a DCCH, and to be paged for a voice call while in service on a PCCH. A conscious decision was made in the 136+ packet data development that voice takes priority over data, so a voice conversation is not interrupted for a packet transaction and simultaneous support for voice and packet data is not required. Paging for one service while in the other service mode requires signaling between the voice and packet data networks. The network architecture selected for the 136+ packet data service allows for this type of operation, and also positions Digital PCS with a graceful path to realize 136HS.

The 136+ packet data network architecture is the same one used for the GSM packet data offering, known as General Packet Radio Service (GPRS) [5]. The GPRS architecture as it applies to the 136+ packet data service is shown in Figure 20.1. The base station subsystem (BSS) is a base station and modified controller to support the 136+ packet data air interface and the Gb' interface to a serving GPRS support node (SGSN). The Gb' interface is a modified version of the GPRS Gb interface to support the 136+ air interface. The SGSN keeps track of the locations of mobiles registered for packet service and performs security functions and access control, in much the same way that the MSC does for mobiles registered for voice service. Signaling between the SGSN and the TIA/EIA-41 MSC passes through an interworking function that translates the GPRS Gs protocol to a TIA/EIA-41 protocol. This signaling is necessary to support voice and data integration, as described previously. The gateway GPRS support node (GGSN) is the interworking function between the local 136+ packet data network and external packet data networks. Communication between the SGSN and GGSN is via the GPRS Gp interface. The GPRS HLR stores subscriber information for packet data service.

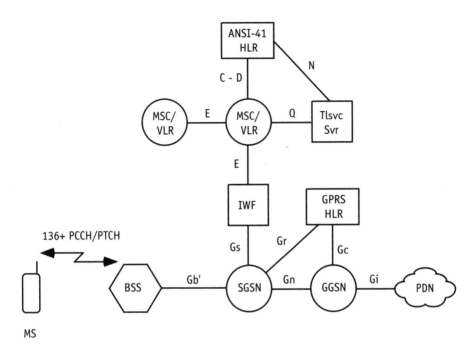

Figure 20.1 136+ packet data network architecture.

Signaling between the GPRS HLR and the SGSN is via the GPRS Gr interface, and signaling between the GPRS HLR and the GGSN is via the GPRS Gc interface.

The packet data network architecture shown in Figure 20.1 is functional, and different network elements could be combined in actual implementations. For example, the GPRS HLR and the TIA/EIA-41 HLR functionality could be combined. The SGSN and the MSC functionality could also be combined. The architecture shown in the figure represents an implementation that requires a minimal amount of change to the existing TIA/EIA-41 network and the GPRS network.

20.3 136HS

The 136HS enhancement was defined to address the IMT-2000 services that cannot be supported on a 30-kHz carrier. Specifically, the IMT-2000 data service objective to provide user data rates up to 384 Kbps is achieved in 136HS by using a 200-kHz wide TDMA channel, and the objective to provide user data rates up to 2,048 Kbps is achieved by using a 1.6-MHz wide TDMA channel. The data service carried on a 200-kHz TDMA channel is compatible with the enhanced data for GSM evolution (EDGE) air interface developed by the European Telecommunications Standards Institute (ETSI). It is also compatible with the GPRS network architecture used with 136+ packet data on a 30-kHz channel.

The definition of 136HS (and 136+ packet data service that uses the GPRS network architecture) increases the commonality between Digital PCS and GSM. This commonality will be important in the future to reduce the total nonrecurring engineering costs associated with the development of the next generation of these systems, to reduce the total bill of materials costs associated with the manufacturing of equipment for these systems, and to increase the potential for roaming among these systems.

20.4 Potential further Digital PCS enhancements

Enhancements to Digital PCS will be continuous in order to stay competitive with other cellular technologies and provide users with the features and functionality they desire. There are two categories of enhancements: those that primarily benefit the user and those that primarily benefit the service provider. Enhancements that benefit the service provider are typically in the area of capacity improvements, which also benefit the user by reducing the blocked call probability. Enhancements to both the 30-kHz and 200-kHz channels can be expected in the future.

The biggest potential voice services enhancement in the near future is the definition of half-rate operation to allow for up to six users per 30-kHz channel. The 8-PSK modulation defined for 136+ enables the development of a half-rate vocoder with voice quality comparable to IS-641-A ACELP. Half-rate operation allows the Digital PCS service provider to obtain approximately twice the current capacity in the same amount of spectrum, an important enhancement as wireless penetration and air time increase.

Potential data services enhancements include the definition of a high-speed circuit-switched capability and simultaneous voice and data capability. However, 136+ does not define a circuit-switched data capability using 8-PSK modulation, and 136HS does not define a circuit-switched data capability for the 200-kHz channel. While packet data fulfills the need for a number of data applications, there are still applications such as file transfer for which circuit-switched connections are more cost-effective. The data rates achievable for a circuit-switched data service using the higher level modulations defined for 136+ and 136HS are the same as those achievable for packet data. It is likely that in the future simultaneous voice and data capability will be achieved, perhaps in both circuit and packet mode. The most likely vehicle for simultaneous voice and data is an enhancement to 136HS, as the 200-kHz channel provides a higher data rate that can be split between the two services. Simultaneous voice and data could also be achieved in the 30-kHz channel through a form of packetized voice, however.

In the realm of teleservices, multimedia capability could be developed for Digital PCS in the future. This would allow for high-quality video imagery and audio to be transmitted across the Digital PCS air interface in support of different mobile applications. Teleservices that allow the user to control his or her feature settings and user profile can also be expected in the future. As applications for the Internet expand, so will the capability of Digital PCS mobiles to take advantage of these applications in the mobile environment.

There will no doubt come a day in the future when Digital PCS is looked upon as the grandfather of a much more sophisticated cellular system, in the same way that AMPS is looked upon today. However, Digital PCS still has a long life ahead of it before that day comes.

References

[1] International Telecommunications Union, *Recommendation ITU-R M.687-2, Future Public Land Mobile Telecommunication Systems (FPLMTS)*, 1997, p. 10.

[2] International Telecommunications Union, *Attachment 4 to Circular Letter 8/LCCE/47*, Apr. 4, 1997, p. 6.

[3] International Telecommunications Union, *Attachment 6 to ITU-R Circular Letter 8/LCCE/47*, Apr. 4, 1997, p. 14.

[4] International Telecommunications Union, *Recommendation ITU-R M.687-2, Future Public Land Mobile Telecommunication Systems (FPLMTS)*, 1997, p. 1.

[5] Cai, J., and D. J. Goodman, "General Packet Radio Service in GSM," *IEEE Communications Magazine*, Vol. 35, No. 10, Oct. 1997, pp. 122–131.

Appendix A
UWCC Member Companies

A.1 Board-Level Companies

AT&T Wireless Services, Inc. (United States)
BellSouth Cellular Corp. (United States)
Cellcom (Israel)
Celumovil (Colombia)
Ericsson Radio Systems
Hughes Network Systems
Lucent Technologies
Mobikom SDN BHD (Malaysia)
Movilnet (Venezuela)
Nokia Mobile Phones
Nortel
Pacific Link Communications Ltd. (Hong Kong)

Philips Consumer Communications
Rogers Cantel, Inc. (Canada)
Southwestern Bell Mobile Systems (United States)
Sun Microsystems Computer Company
Tandem Computers, Inc.
Telecom New Zealand, Ltd. (New Zealand)
Uniden
Vimpelcom (Russia)

A.2 General Membership

3dbm, Inc.
AG Communication Systems
Aldiscon
Antel-Ancel (Uruguay)
Allen Telecom, Inc.
Astronet Corporation
AVAL Communications, Inc.
Bellcore
Comcast Cellular Communications, Inc. (United States)
Comcel S.A. (Colombia)
Conecel (Ecuador)
Dobson Cellular (United States)
Domital Corporation
EDS Personal Communications
Ghuangzhou CESEC (China)
Global Mobility Systems
Harris Canada, Inc.
IBM
IFR Systems, Inc.
Indus, Inc. (United States)
ISOTEL Research
Iwix Communications, Inc.
Kokusai Communications Systems
MCOMCAST S.A. (Brazil)
Mercury PCS, LLC (United States)

Miniphone S.A. (Argentina)
Mitsubishi
Motorola
Natural Microsystems
NEC America Mobile Phones
Octel Communications
Otecel S.A. (Ecuador)
Powerwave Technologies
PriCellular Corp. (United States)
Racotek-NextNet
RANN International (United States)
Ratelindo (Indonesia)
SEMA Group Telecoms
SNET Mobility (United States)
Sony Corporation
SPT Telecom (Czech Republic)
Startel S.A. (Chile)
TATINCOM, Tatarstan International Communications, Ltd. (Russia)
Telcel (Mexico)
Telecom Personal S.A. (Argentina)
Telecom Personal S.A. (Paraguay)
Telecorp (United States)
Tellabs Wireless International
Telogy Networks
Telos Engineering Limited
Texas Instruments Wireless Communications
TriQuint Semiconductor
Triton Communications (United States)
Unifon (Argentina)
Unwired Planet
Vanguard Cellular Systems, Inc. (United States)
Watkins-Johnson Company
Wink Communications
Wireless One (United States)

Glossary

A-Key
A 64-bit secret key known only to the mobile and the home authentication center and used to generate shared secret data to support the authentication, privacy, and encryption algorithms.

abbreviated burst
A shortened slot format on the reverse digital control channel used for large cell sites to ensure that mobiles transmitting on adjacent time slots do not overlap each other when one mobile is located at the cell boundary and the other mobile is located close to the base station.

acceptable service provider
A home or partner service provider. When a mobile finds a suitable control channel of an acceptable service provider while executing intelligent roaming scanning procedures, it proceeds to obtain service and does not execute triggered scans.

access response channel (ARCH)
A logical channel of the SPACH used to deliver point-to-point responses to messages from mobiles.

adaptive channel allocation (ACA)
A method of increasing the capacity of a cellular system while maintaining the engineered grade of service by automatically and dynamically changing the frequency reuse pattern to optimally match the propagation environment and traffic distribution.

Advanced Mobile Phone Service (AMPS)
A cellular radio telephone system that uses Frequency Division Multiple Access (FDMA) in a cell site and frequency reuse among cell sites to maximize the number of simultaneous users in a given spectrum allocation.

algebraic code excited linear predictive coding (ACELP)
A form of source coding used to transmit voice over a DTC in Digital PCS.

alpha tag
The system name that may be displayed to the user while a mobile is operating on a control channel. The alpha tag may be up to 15 characters in length.

American National Standards Institute (ANSI)
The primary standards making group in the United States responsible formulating and accrediting telecommunications standards.

analog control channel (ACC)
A 30-kHz channel using frequency shift keying (FSK) modulation to convey signaling information between a mobile and the network at a rate of 10 Kbps.

analog voice channel (AVC)
A 30-kHz channel using frequency modulation (FM) on which a voice conversation occurs and on which brief digital messages may be sent between a mobile and the network.

automatic retransmission request (ARQ)
A mode of operation used for acknowledged delivery of layer 3 information to a mobile.

authentication
The process of verifying the authenticity of a mobile.

authentication center (AC)
The cellular network element responsible for managing authentication information related to a mobile, including the A-Key and SSD for a mobile. It also executes the CAVE algorithm to generate SSD, perform authentication, and calculate the voice privacy mask and signaling message encryption key for a particular mobile.

autonomous system
A microsystem of network type private and/or residential that shares frequencies with the cellular network.

band history table (BHT)
A dynamic database stored in the mobile that is used to speed the process of finding a control channel of an acceptable service provider when conducting a triggered scan from an unacceptable service provider.

bandpass filter (BPF)
A filter that passes frequencies within the passband and attenuates frequencies above and below this region.

base station (BS)
A cellular network element composed of a radio transceiver controller, one or more radio transceivers, and an antenna system, used to communicate with mobiles stations over the air.

base station manufacturers code (BSMC)
An 8-bit value broadcast on the DCCH to identify the manufacturer of the network infrastructure. TIA/EIA-136 allows for the development of BSMC-specific signaling.

base station, mobile switching center, and interworking function (BMI)
A network entity in a simple Digital PCS network reference model that communicates with a mobile, other BMIs, and the PSTN.

bit error rate (BER)
For a given digital modulation technique, the probability of bit error as a function of carrier-to-noise ratio.

broadcast control channel (BCCH)
A shared, point-to-multipoint, unacknowledged logical channel on the forward DCCH.

camping
Short for the DCCH Camping state, the state in which a mobile is idle on a DCCH and available for service.

Candidate Eligibility Filtering (CEF)
A part of the Reselection procedure that a mobile executes while operating on a DCCH to screen neighboring control channels for further examination.

carrier-to-interference ratio (C/I or CIR)
The ratio of carrier signal power to interference signal power, with decibel (dB) as the typical unit of measure.

cell site
The physical location of a base station that serves a geographic area in a cellular network.

CELLTYPE
An information element broadcast on the DCCH that identifies whether the DCCH is serving a Regular, Preferred, or Nonpreferred sector of a cell. The mobile uses CELLTYPE in the reselection procedure to prioritize neighboring control channels.

cellular authentication and voice encryption (CAVE)
A randomization algorithm used to calculate shared secret data, authentication responses, voice privacy masks, and signaling message encryption keys.

cellular messaging teleservice (CMT)
A teleservice for the delivery of alphanumeric messages, or short messages, between two short message entities.

Cellular Telecommunications Industry Association (CTIA)
An international wireless communications industry association of wireless service providers and manufacturers.

channel quality measurement (CQM)
The signal-quality measurements and subsequent reports made by a mobile as part of MAHO. CQM reports include the signal strength on the current channel and on handoff candidates. CQM reports also include error rate measurements on the current channel, and may include a C/I estimate as well.

closed user group
A nonpublic service feature that allows certain cell sites to only be accessed by mobiles that subscribe to the network type and PSID or RSID broadcast on the DCCH.

coded DCCH locator (CDL)
An encoded field on a digital traffic channel indicating a range of channels in which a digital control channel may be found.

coded DVCC (CDVCC)
An encoded field on the digital traffic channel and digital control channel containing a digital verification color code used to indicate that the correct rather than cochannel data is being decoded.

comfort noise (CN)
Artificial noise inserted into a conversation when there is no voice activity in order to produce more realistic background noise than is otherwise possible when a vocoder is used to digitize one or both of the callers' speech.

control channel
Either an ACC or DCCH used to communicate signaling information between a mobile and the network.

control mobile attenuation code (CMAC)
A 3-bit field indicating the power level for a mobile to use when accessing an ACC.

convolutional code
A type of channel encoding that operates on a span of information bits that is shifted one information bit every time another information bit is input to the encoder.

cyclic redundancy check (CRC)
A parity-check code capable of detecting multiple errors in a block of information bits.

DCCH history table (DHT)
A dynamic database stored in the mobile that is used to speed the process of finding a DCCH upon power-up.

differential quadrature phase shift keying (DQPSK)
The type of digital modulation used on a digital control channel and digital traffic channel.

digital color code (DCC)
A digital signal transmitted by the base station on a forward analog control channel that is used to detect capture of a base station by an interfering mobile station.

digital control channel (DCCH)
A full-rate or half-rate logical channel on a 30-kHz RF channel using $\pi/4$ DQPSK modulation to convey signaling information between a mobile and the network.

digital control channel locator (DL)
A 7-bit data field on the DTC sent by the base station to help a mobile find a DCCH. It contains a range of 8 or 16 channels in which a DCCH is located.

digital traffic channel (DTC)
A full-rate or half-rate logical channel on a 30-kHz RF channel using $\pi/4$ DQPSK modulation to convey a voice conversation, data call, or digital messages between a mobile and the network.

digital mobile attenuation code (DMAC)
A 4-bit field indicating the initial power level for a mobile to use upon assignment to a DTC.

digital verification color code (DVCC)
An 8-bit code sent by the base station to the mobile used to indicate that the correct rather than cochannel data is being decoded.

directed retry
The process of directing a mobile to reoriginate or resend a page response on a different control channel due to network congestion or quality of service issues.

discontinuous transmission (DTX)
A mode of operation for a mobile while in the conversation state in which the mobile's transmitter is switched on an off based on voice activity detection, leading to greater talk time.

downlink power control (DPC)
A feature that allows a base station to transmit at different power levels on each time slot of an RF channel. Downlink power control can reduce the interference level caused by base station transmissions and lead to tighter frequency reuse.

effective radiated power (ERP)
The combination of transmitter power and antenna gain referenced to a half-wave dipole.

electronic data interchange (EDI)
A method of data transfer used to communicate ESN/A-Key pairs between a mobile manufacturer and a service provider.

Electronics Industry Association (EIA)
A standards-making organization in the United States that covers all areas of electronics information and communications technology.

electronic serial number (ESN)
A 32-bit number stored permanently in a mobile used to uniquely identify the equipment.

extended broadcast control channel (E-BCCH)
A logical channel of the BCCH used to carry less time-critical information than the F-BCCH.

fast associated control channel (FACCH)
A blank-and-burst channel on the digital traffic channel used for signaling exchange between the mobile and base station.

fast broadcast control channel (F-BCCH)
A logical channel of the BCCH used to carry time-critical information essential for mobiles to access the network.

favored service provider
A service provider with a favored SID, favored SOC, or both, contained in the IRDB of a mobile. A favored service provider is of higher priority than any other service provider except the home and partner service providers.

first office application (FOA)
A technical trial of a new feature or service, typically with live users in a commercial market.

forbidden service provider
A service provider with a forbidden SID, forbidden SOC, or both, contained in the IRDB of a mobile. The mobile can only obtain emergency service from a forbidden service provider.

forward analog control channel (FOCC)
An analog control channel used from the base station to the mobile.

forward digital traffic channel (FDTC)
A digital traffic channel used from the base station to the mobile.

forward voice channel (FVC)
An analog voice channel used from the base station to the mobile.

frame error rate (FER)
The ratio of incorrectly received frames to total frames transmitted, where an incorrectly received frame is defined as a frame of data having a CRC check that fails.

frequency division duplex (FDD)
A method for allowing a user to send and receive information at the same time by using different frequencies for transmission and reception.

Frequency Division Multiple Access (FDMA)
A method for dividing the available frequency spectrum into channels and assigning the channels to different users to allow for simultaneous use of the frequency spectrum.

frequency shift keying (FSK)
A type of digital modulation in which information is conveyed in the frequency change of the carrier.

general availability (GA)
The commercial availability of a new feature or service.

general UDP transport teleservice (GUTS)
A general-purpose application data delivery service that provides an open platform for the creation of wireless information services.

Global Operators' Forum (GOF)
A technical arm of the UWCC responsible for the development of common service provider requirements and operational standards and procedures.

Global System for Mobile Communications (GSM)
A second-generation digital cellular system using TDMA with 200-kHz channels and GMSK modulation. In North America, GSM has been up-banded to work in the 1,900-MHz frequency band and is known as PCS1900.

Global TDMA Forum (GTF)
A technical arm of the UWCC that focuses on the evolution of the TIA/EIA-136 air interface for Digital PCS.

Global WIN Forum (GWF)
A technical arm of the UWCC that focuses on developing new wireless intelligent network (WIN) features for Digital PCS based on the TIA/EIA-41 intersystem standard.

handoff
The process of transferring a mobile from one voice or traffic channel to another voice or traffic channel to maintain a desired quality of service level for the mobile user.

hierarchical cell structures (HCS)
A capability enabled by Digital PCS to define layers of service to increase the capacity of a cellular system. This layering is accomplished by defining cells or sectors with a CELLTYPE of Preferred, Regular, or Nonpreferred, and by changing the delay associated with reselecting to different DCCHs.

higher layer protocol identifier (HLPI)
An 8-bit field that identifies the teleservice included in a data unit.

historic search
A portion of the intelligent roaming scanning procedure in which a mobile searches for previously used DCCHs.

home location register (HLR)
The primary database in the cellular network that stores information about mobile users. It contains a record for each home subscriber that includes the mobile directory number, subscriber features, subscriber status, and current location information.

home service provider
A service provider broadcasting a SID and/or SOC that matches the home SID and/or SOC stored in the mobile.

hyperband
Either the 800-MHz or 1,900-MHz frequency band.

hyperframe (HF)
A measure of time on the DCCH consisting of 32 TDMA frames (1.28s), or two superframes.

information element (IE)
A portion of layer 3 data that contains a particular type of information.

intelligent roaming (IR)
A method for ensuring that a mobile obtains service from the best provider in an area without requiring user intervention.

intelligent roaming database (IRDB)
A database of roaming information stored in a mobile and used to execute the intelligent roaming procedures. Among other things, the IRDB contains the band order for scanning for service and lists of partner, favored, and forbidden service providers.

interactive voice response (IVR)
A feature available with some peripheral equipment, such as voice mail, that prompts the caller to enter information and act upon that information.

Interim Standard (IS)
A type of TIA standard released for industry use for a limited period of time, typically three years.

international mobile station identity (IMSI)
A 15-digit address that uniquely identifies the mobile, the home wireless network, and the home country of the network and the mobile.

International Organization for Standardization (ISO)
A voluntary, nontreaty organization that develops international standards.

Interoperability Test Group (ITG)
A working group of the Global TDMA Forum that develops test specifications for new Digital PCS features and services.

interworking function (IWF)
A cellular network element that performs protocol conversion operations.

layer
A plane in the OSI reference model that performs a specific function and depends upon the layer below it to provide services.

layer 1
The physical layer in the OSI reference model, providing a mechanism to send and receive a stream of bits over a physical transmission medium.

layer 2
The data link layer in the OSI reference model, providing addressing; frame delimiting; error detection, recovery, and sequencing; media access control; and flow control of data transferred between network layer entities.

layer 3
The network layer in the OSI reference model, providing a means to establish, maintain, and terminate connections between end systems.

low-noise amplifier (LNA)
An amplifier with a low-noise figure placed in the receive chain to increase the sensitivity of the receiver.

message center (MC)
The same as a short message service center.

message encryption algorithm (MEA)
A 3-bit field that indicates the message encryption algorithm to be used.

message encryption key (MEK)
A 3-bit field that indicates the message encryption key to be used.

message transfer part (MTP)
A portion of the TIA/EIA-41 SS7-based data transfer service that describes the physical, data link, and part of the network layer of SS7.

Message Type (MT)
The second information element in a DCCH layer 3 message, defining how to interpret the remaining information elements.

message waiting indicator (MWI)
A notification provided to the mobile user through network signaling to indicate that one or more messages (typically voice messages) have been received.

mobile application part (MAP)
A portion of the TIA/EIA-41 application services that provides the mobility management function.

mobile assisted channel allocation (MACA)
A procedure in which the network requests mobiles operating on a DCCH to measure and report the signal strength on up to 15 channels in addition to the current DCCH. The MACA report provides the network with downlink signal strength measurements on channels that may be in use in other sectors.

mobile assisted handoff (MAHO)
A process in which a mobile operating on a DTC measures signal quality on the current DTC and other RF channels specified by the network and reports this information to the network. The network uses the signal-quality information reported by the mobile in conjunction with measurements made by the network to decide if the mobile should be handed off to another voice or traffic channel.

mobile country code (MCC)
A 3-digit decimal digit coded as 10 bits and used to identify a country.

mobile identification number (MIN)
The 34-bit representation of the 10 digit directory number assigned to a mobile.

mobile station (MS)
A cellular radiotelephone.

mobile station identity (MSID)
A permanent or temporary address used identify a mobile over the air interface.

mobile station power class
A rating for the maximum effective radiated power of a mobile.

mobile switching center (MSC)
A cellular network element that provides the same functionality as a central office switch in the public switched telephone network and is addi-

tionally responsible for call processing, mobility management, and radio resource management.

neighbor list
A listing of the control channels in neighboring cells and sectors that is broadcast on a DCCH and used by mobiles as candidates for reselection.

network type
An identification of the type of service that is offered on the system. The network type is broadcast on the DCCH and may be public, private, and/or residential.

neutral service provider
A service provider whose SID and SOC are not contained in a mobile's IRDB. A mobile may obtain service on a suitable control channel of a neutral service provider if no higher priority service providers are available in the area. A mobile executes triggered scans from a neutral service provider in an attempt to find a control channel of a higher priority service provider.

nonpublic service
Service offered from a system whose DCCH indicates a network type of private and/or residential and that broadcasts PSIDs and/or RSIDs upon which one or more mobiles may obtain service, thereby forming a closed user group.

normal burst
A normal length frame on the reverse DCCH and reverse DTC.

numeric assignment module (NAM)
A portion of a mobile's nonvolatile memory containing subscriber and home system information, typically programmed by the home service provider at the time of activation.

Open Systems Interconnection (OSI)
A seven-layer reference model designed by the ITU to facilitate exchange of information between systems. The seven layers are physical, data link, network, transport, session, presentation, and application.

operations, administration, and maintenance (OA&M)
A function that allow the service provider to control the operation of the cellular network, extract billing and traffic data, and detect and recover from faults.

origination
The process of originating a call from a mobile when the mobile is operating on a control channel.

over-the-air activation (OAA)
The overall process of programming a mobile's NAM over the air.

over-the-air activation function (OTAF)
A functional entity within a teleservice server that formats, stores, and forwards over-the-air activation messages.

over-the-air activation teleservice (OATS)
A teleservice used in over-the-air activation to program the NAM and A-Key of a mobile.

over-the-air programming (OAP)
The overall process of programming non-NAM data into a mobile over the air.

over-the-air programming teleservice (OPTS)
A teleservice used to program a mobile's IRDB.

packet control channel (PCCH)
A logical channel, similar in function to a DCCH, upon which a mobile camps while awaiting a packet transaction. The PCCH may also carry short packet transactions.

packet traffic channel (PTCH)
A logical channel, similar in function to a DTC, upon which a mobile transmits and receives data packets.

paging channel (PCH)
A logical channel on the SPACH used to deliver pages and orders to mobiles.

paging frame class (PFC)
The interval in hyperframes between occurrences a mobile's PCH.

partner service provider
A service provider defined by a partner SID, partner SOC, or both, contained in the IRDB of a mobile. A partner service provider is of higher priority than any other service provider except the home service provider. A mobile does not conduct triggered scans from a partner service provider.

permanent mobile station identity (PMSID)
A permanent address stored in the mobile's NAM and used to identify the mobile over-the-air interface, either the MIN or IMSI.

personal communications service (PCS)
A set of advanced wireless services that may be offered to a mobile user through the cellular network, regardless of the frequency band.

phase shift keying (PSK)
A type of digital modulation in which information is conveyed in the phase change of the carrier.

physical layer control
A message sent by the network to a mobile to command a change in certain mobile parameters such as power output and time alignment.

Power Amplifier (PA)
A device that amplifies the RF output of a modulator for transmission.

Priority System condition
A Reselection Trigger Condition (RTC) invoked when a mobile finds one or more control channels matching a public service profile (PSP) stored for an autonomous system with which it has subscribed. When this RTC is invoked, the mobile measures the signal strength on the POFs that correspond to the PSP, then executes Candidate Eligibility Filtering in an attempt to reselect to the autonomous system if it is suitable for service.

private and residential system identities (PSID/RSID)
A 16-bit value used to identify a private or residential system. A mobile obtains service on the private or residential system by matching the

broadcast PSID or RSID with an internally stored value, then registering with the PSID or RSID.

private operating frequency (POF)
Frequencies that are currently used, or may have a high probability of being used, as one or more DCCHs for an autonomous system. A mobile stores POFs for autonomous systems with which it has subscribed and checks these frequencies upon power-up and upon declaring a Priority System condition.

probability block
A range of RF channels in the 800-MHz band that has a relative probability associated with finding a DCCH within the range.

protocol
A set of rules and formats for communications between peer entities to provide services or implement functions.

Protocol Discriminator (PD)
The first information element in a DCCH layer 3 message, identifying the standard that defines the message.

public service profile (PSP)
The frequency, DVCC or DCC, and SID of a control channel in the public cellular network that is stored in a mobile and used to trigger reselection to an autonomous system.

public switched telephone network (PSTN)
The telecommunications network that provides telephone service to the general public.

Radio Link Protocol 1 (RLP1)
A medium access control layer defined for circuit-switched data on a DTC.

random access channel (RACH)
A shared, point-to-point, acknowledged logical channel used by mobiles to access the digital control channel.

received signal strength (RSS)
The signal strength measured by either the mobile or base station, typically described in units of decibels referred to a milliwatt, or dBm.

registration
The process whereby a mobile identifies itself to the network and makes itself available for service.

reselection
The process whereby a mobile that is operating on a DCCH selects the best neighboring control channel based on signal strength measurements, trigger conditions, and parameters settable by the service provider.

Reselection Trigger Condition (RTC)
A criteria that causes a mobile to invoke the Candidate Eligibility Filtering procedure as part of reselection. The RTCs are Radio Link Failure, Cell Barred, Server Degradation, Directed Retry, Priority System, Service Offering, and Periodic Evaluation.

reverse analog control channel (RECC)
An analog control channel used from the mobile to the base station.

reverse analog voice channel (RVC)
An analog voice channel used from the mobile to the base station.

reverse digital traffic channel (RDTC)
A digital traffic channel used from the mobile to the base station.

service access point (SAP)
A point at which service is provided by one layer to the next highest layer.

service primitive
An implementation-independent interaction between two layers at a service access point.

service provider (SP)
A carrier that operates a cellular network and is licensed to provide service in a geographic area.

Glossary **361**

Shared Channel Feedback (SCF)
Feedback used to control mobile access on the reverse DCCH with three subfields: Busy/Reserved/Idle (BRI), Coded Partial Echo (CPE), and Received/Not Received (R/N).

shared secret data (SSD)
A 128-bit pattern stored in a mobile's semipermanent memory and in the home authentication center. The first 64-bit group of the SSD is called SSD-A and is used to support authentication. The second 6-bit group of the SSD is called SSD-B and is used to support voice privacy and message encryption.

short message entity (SME)
A functional network entity that can originate and terminate short messages.

short message service (SMS)
The same as cellular messaging teleservice.

SMS-broadcast control channel (S-BCCH)
A logical channel of the BCCH used to carry broadcast teleservices.

short message service center (SMSC)
A functional entity within a teleservice server that formats, stores, and forwards short messages.

SMS channel (SMSCH)
A logical channel of the SPACH used to deliver teleservice-related messages to mobiles.

short message terminal (SMT)
The portion of a mobile station that encodes and decodes short messages.

shortened burst
A shortened slot format on the reverse DTC sometimes required of a mobile to avoid collisions at the base station between the mobile's burst and the burst of a neighboring slot.

signal-to-noise ratio (SNR or S/N)
The ratio of signal power to noise power, with decibel (dB) as the typical unit of measure.

signaling connection control part (SCCP)
A portion of the TIA/EIA-41 SS7-based data transfer service that describes part of the network layer of SS7.

signaling message encryption (SME)
A feature available on both the analog voice channel and digital traffic channel in which a limited number of messages containing sensitive subscriber information are encrypted prior to transmission over the air and decrypted at the receiver.

Signaling System 7 (SS7)
A suite of common channel network signaling protocols defined for North American telecommunications networks, and one of the protocols that may be used with TIA/EIA-41.

sleep mode
The state in which a mobile switches off devices for a period of time when they are not needed in order to increase standby time.

slow associated control channel (SACCH)
A continuous channel on the digital traffic channel used for signaling message exchange.

SMS point-to-point, paging, and access response control channel (SPACH)
A shared, point-to-point channel on the forward DCCH that is further divided into three logical channels: SMSCH, PCH, and ARCH.

standby time
A measure of the total time a mobile may be idle on a control channel before the mobile's battery needs recharging.

sub-band
A portion of a 1,900-MHz band, typically 83 channels.

superframe (SF)
A measure of time on the DCCH consisting of 16 TDMA frames, or 640 ms.

supervisory audio tone (SAT)
One of three tones in the 6-kHz region that are transmitted by a base station and transponded by a mobile.

system identity (SID)
A 15-bit value used to uniquely identify a licensed service provider within a particular geographic area.

system operator code (SOC)
A 12-bit value used to uniquely identify a licensed service provider without respect to a particular geographic area.

talk time
A measure of the total time a mobile may be in the conversation state on an AVC or DTC before the mobile's battery needs recharging.

Telecommunications Industry Association (TIA)
A trade organization that operates in conjunction with the Electronics Industry Association to develop telecommunications standards for North America.

teleservice
An application that uses the air interface and network interface as a bearer for transport between a network element called a teleservice server and a mobile.

teleservice segmentation and reassembly (TSAR)
A transport layer service that allows long teleservice messages to be broken into smaller segments for transmission through the network and across the air interface, then put back together at the receiving end.

teleservice server
A cellular network element that formats, stores, and forwards teleservice layer messages to and from a mobile.

temporary mobile station identity (TMSI)
A 20- or 24-bit address assigned to a mobile at registration and used to identify the mobile over the air for efficient paging and fraud prevention.

termination
The process of a mobile receiving a call.

thin client architecture (TCA)
An architecture including intelligent mobiles, the cellular network, the Internet, and World Wide Web sites to deliver text-based information services to a mobile user.

time alignment
A process that allows for the advancing of the offset between the reverse and forward frames of a digital traffic channel by up to 15 symbols, in half-symbol units, thereby allowing a mobile operating far from the base station to transmit in a time slot without interfering with a mobile operating close to the base station and transmitting in the next time slot.

time division duplex (TDD)
A method for allowing a user to send and receive information on the same RF channel by splitting the channel into segments of time for transmission and different segments for reception.

Time Division Multiple Access (TDMA)
A method for allowing more than one use of a physical channel for different purposes by sequentially assigning the channel to different users for a specified amount of time.

transaction capabilities application part (TCAP)
A portion of the TIA/EIA-41 SS7 application services responsible for information transfer between two or more nodes in the signaling network.

unacceptable service provider
A favored, neutral, or forbidden service provider.

Universal Wireless Communications Consortium (UWCC)
A cellular industry organization consisting of leading Digital PCS vendors and service providers.

vector sum excited linear predictive coding (VSELP)
An early vocoder used in IS-54 systems.

visitor location register (VLR)
A database function local to an MSC that maintains temporary records associated with nonhome mobiles registered for service.

voice mobile attenuation code (VMAC)
A 3-bit field indicating the initial power level for a mobile to use upon assignment to an AVC.

voice privacy (VP)
A feature available with digital voice coding in which the encoded speech is encrypted prior to transmission over the air and decrypted at the receiver to provide more private conversations.

voltage-controlled oscillator (VCO)
A sine wave generator with a variable frequency.

wireless intelligent network (WIN)
A suite of custom features that a service provider may implement without the need for individual feature standardization, including voice-controlled services, incoming call screening, and calling name identification presentation.

word error rate (WER)
The same as frame error rate.

About the Author

CAMERON COURSEY is a Member of Technical Staff at SBC Technology Resources, Inc., the research and development arm of SBC Communications, Inc. (SBC). He has worked in the wireless industry since 1991. Since 1994, Mr. Coursey has been involved in the development of Digital PCS features and services for SBC. His primary activities have been the development of technical specifications for Digital PCS mobiles, acceptance testing of Digital PCS mobiles, development of software programs for Digital PCS network testing and optimization, development and implementation of test plans for the acceptance testing of Digital PCS mobiles and networks, development and standardization of intelligent roaming algorithms and nonpublic services, representation of SBC in the Global TDMA Forum (GTF) of the Universal Wireless Communications Consortium (UWCC), chairmanship of the 136+ Working Group of the GTF directing the development of requirements for the next generation of the Digital PCS voice and data services, and representation of SBC in the Telecommunications Industry Association

(TIA) TR-45.3 engineering subcommittee, the Digital PCS standards development organization.

Mr. Coursey obtained his Master of Science degree in Electrical Engineering from the University of Missouri-Rolla in 1988, and his Bachelor of Science in Electrical Engineering a year earlier from the same institute. Prior to joining SBC, Mr. Coursey worked for three years developing advanced communications systems for military aircraft at McDonnell-Douglas Corporation.

Index

π/4 shifted, differentially encoded
    quadrature phase shift keying
    (π/4 DQPSK), 51
136+, 13–14, 39, 41, 333–35
    packet data network
        architecture, 334–35
    packet data service, 333
    voice/data service integration, 334
    voice services, 333
    Working Group, 327, 328
    See also IS-136
136HS, 14–15, 335–36
    air interfaces, 14–15
    defined, 14, 335–36
    definition of, 336
    See also IS-136

Access attempts, 299–302
Access Parameters message, 92–93
Access response channel (ARCH), 62
Access thresholds, 296–99
    defined, 296
    MS_ACC_PWR, 296–99
    RSS_ACC_MIN, 296–99
Adaptive channel allocation (ACA), 20
    capacity increase with, 29
    defined, 27–28
    with HCS, 28
Adaptive differential pulse-code
    modulation (ADPCM), 332
Advanced Mobile Phone Service
    (AMPS), 19
    defined, 1
    Digital PCS compatibility, 33–34
    feature comparison, 10
    finding ACC, 3
    handoff algorithm, 5
    IS-54 channel comparison, 7
    network illustration, 5
    network migration, 33
A interface, 116
A-key, 8, 226, 274

A-key (continued)
    defined, 274
    generation, 275
    transfer, 275
Alert with Info message, 100–101
Algebraic code excited linear predictive
        coding (ACELP), 10, 186
American National Standard, 38
Analog circuit-switched data
        service, 231–33
    architecture, 232
    support, 231–32
Analog control channels
        (ACCs), 2, 11, 129
    AMPS mobile finding, 3
    PSP characteristics, 251
Analog-to-digital (A/D) conversion, 185
Analog voice channel (AVC), 5
    assignment, 4
    assignment message, 4
    channel number, 95
Analog Voice Channel Designation
        message, 95
Antennas, smart, 111
ARQ mode, 85
Asynchronous data and group 3 fax
        service, 233, 240
Audit message, 101
    Confirmation, 98
    Order, 95
Authentication, 8
    global challenge, 135, 276–78
    key, 16
    procedures, 135, 275–80
    unique challenge, 278–80
Authentication center (AC), 108, 114
    data link, 114
    defined, 114
Authentication message, 98
Autonomous systems, 249–55
    advantages, 249–50
    cell sites, 250
    DCCHs, 250, 251
    defined, 249
    Digital PCS and, 250
    example of, 251–52

Band history table (BHT), 267
Base Station message
    Ack, 101
    Challenge Order, 99, 103
    Challenge Order Confirmation, 95, 101
Base stations (BSs), 4, 5, 6, 55,
        107, 110–12
    antenna system, 111
    components, 110
    power output requirements, 56
    radio transceiver controller, 110
    residential, 249–50
BCCH messages, 92–95
BMI
    defined, 107
    model, 108
Broadcast change notification
        (BCN), 179, 180
Broadcast control channel (BCCH), 62
    Access Parameters, 92–93
    changes to, 179, 180
    Control Channel Selection
        Parameters, 93
    DCCH Structure, 92
    Mobile Assisted Channel
        Allocation, 94–95
    Neighbor Cell, 94, 142, 143
    Registration Parameters, 93
    Regulatory Configuration, 94
    rereading, 179–81
    Service Menu, 93–94
    sleep mode and, 179–81
    SOC/BSMC Identification, 94
    System Identity, 93
    See also BCCH messages
Broadcast teleservice transport, 213–17
Broadcast teleservice transport information
        (BTTI), 214–16
    defined, 214
    parameters, 215
    summary, 216
Bursts, 301–2

Calling number identification, 8
Call processing, 194–203
    data, 240–41
    defined, 194

establishment, 194–99
functions, 112
handoff, 199–201
termination, 197
vocoder assignment, 201–3
Candidate Eligibility Filtering (CEF)
    procedure, 146–55
    calculations, 153
    CEF1, 147, 148
    CEF2, 147–49, 151
    CEF3, 150, 151–53
    CEF4, 151, 153–55
    CEF5, 152, 154–55, 167
    defined, 146
    mapping of RTCs into, 147
    parameters, 146, 153
Candidate Reselection Rules
    procedure, 155–57
    CAND_1 Determination, 155–56
    CAND_1 Examination, 156
    Suitable CAND_1 Found, 157
    Suitable CAND_1 not Found, 157
Capability Report message, 99
Capability Update Request
    message, 101, 103
Capability Update Response
    message, 101, 103
Capacity
    calculating, 23
    instantaneous, 22
Capacity enhancement, 20–29
    with ACA, 27–29
    with HCS, 24–27
    with TDMA, 20–24
    See also Digital PCS
CAVE algorithm, 127, 273–74, 284
    defined, 273–74
    inputs, 274
    running of, 274
Cell Barred example, 159–61
    CEF2 calculations for, 160
    defined, 159
    See also Reselection
Cellular messaging teleservice
    (CMT), 115, 218–23
    defined, 12, 218
    SMS DELIVER message, 218–21

SMS DELIVERY ACK message, 218
SMS MANUAL ACK message, 218
SMS SUBMIT message, 218–21
uses, 218
See also Teleservices
Cellular Telecommunications Industry
    Association (CTIA), 322–23
    certification, granting, 323
    certification seal, 322
    testing, 322, 323
Channel coding, 66–69
    DCCH, 68–69
    defined, 66
    DTC, 69
    IS-641, 188–89, 190
Channelization, 48–50
    in 800-MHz hyperband, 49
    in 1900-MHz hyperband, 50
Channel Quality Message message, 103
Circuit-switched data services, 231–41
    analog, 231–33
    digital, 233–41
Cloning fraud, 275
Closed-loop pitch analysis, 188
Closed user groups, 244
    defining, 245
    forming, illustrated, 246
Coded DCCH locator (CDL), 296
Coded DVCC (CDVCC), 88
Code excited linear prediction
    (CELP), 186–87
    algebraic (ACELP), 10, 186
    speech-generation model, 187
Comfort noise (CN), 42, 191–93
Connect message, 103
Consolidated Protocol Test Specification, 318
Control Channel Locking
    procedure, 143–44
Control Channel Reselection
    procedure, 130–31
    Control Channel Locking, 130–31
    execution of, 130
    Reselection Criteria, 131
Control Channel Scanning and Locking
    state, 120–22
Control Channel Selection Parameters
    message, 93

Control Channel Selection
procedure, 129–30
execution of, 129
parts, 129
See also Mobile Station procedures
Control mobile attenuation code
(CMAC), 299
Convolutional code, 68
Cyclic redundancy check (CRC) code, 68

Data communication equipment
(DCE), 235
Data link layer, 71–88
DCCH, 73–87
defined, 30, 71
DTC, 87–88
E-BCCH frames, 78–79
F-BCCH frames, 78–79
RACH frames, 77, 78
SPACH frames, 80–82
See also Layered protocol
Data terminal equipment (DTE), 233
AT command entry, 240
far-end, 235
MT communication with, 239
DCCH, 22
ACC vs., 11
autonomous system, 250, 251
channel coding, 68–69
composition, 11
data link layer, 73–87
defined, 10
encryption on, 284–85
in forward/reverse direction, 59, 62
full-rate, 67
half-rate, 57
history table (DHT), 261
layer 2
frame structure, 77
functions, 74
media access control, 82–85
protocols, 76–82
location, 291–96
logical channels, 62
message set, 92–100
neighbor list, 11
optional information elements, 90

paging frame class (PFC), 11
physical layer, 48
radio link quality, monitoring, 87
service access points, 73–76
structure, 11, 302–3
teleservices, 12, 15
time-slot fields, 61
time-slot formats, 60
time-slot structure, 59–63
DCCH Camping state, 122–23, 202
DCCH Scanning and Locking
procedure, 128–29
DCCH Structure message, 92
DCS1800. *See* GSM
Dedicated DTC Handoff message, 101
Defined teleservices, 206, 217–30
cellular message (CMT), 218–23
defined, 206
generic UDP transport
(GUTS), 14, 229–30
over-the-air activation (OATS), 223–27
over-the-air programming
(OPTS), 227–28
See also Teleservices
Differentially encoded quadrature phase
shift keying (DQPSK), 51
modulator, 51
phase change mapping, 52
phase constellation, 54
Digital circuit-switched data
service, 233–41
defined, 233
network architecture, 234
protocol stack, 234
Digital control channel. *See* DCCH
Digital mobile attenuation code
(DMAC), 299
Digital PCS
advantages, 19–34
AMPS compatibility, 33–34
autonomous systems and, 250
battery life, 11
capacity enhancement, 20–29
defined, xiii
enhancements, potential, 336–37
feature comparison, 10
feature flexibility, 29–32

future of, 331–37
genealogy, 16–17
GSM relationship with, 15–16
layered protocol, 29–31
mobile operation, 119–39
MWI, 12
network support, 12
standards, 35–45
standby time, 12
Digital signal processors (DSPs), 30, 193
Digital traffic channel (DTC), 6
  assignment, 7, 8
  channel coding, 69
  channel number, 95
  data link layer, 87–88
  messages, 100–105
  physical layer, 48
  shortened burst on, 65
  time-slot fields, 65
  time-slot formats, 64
  time-slot structure, 63–66
Digital Traffic Channel Designation
    message, 95
Digital verification color code
    (DVCC), 87–88, 92, 296
  characteristics, 88
  check, 87
  coded (CDVCC), 88
  defined, 87
Directed Retry example, 162–63
  CEF3 calculations for, 163
  defined, 162
  See also Reselection
Directed Retry message, 96, 196, 198
Discontinuous transmission
    (DTX), 42, 191–93
Dual-band mobiles, 270, 271
Dual-mode mobiles, 270, 271
Duplexing, 48–50
  in 800-MHz hyperband, 49
  in 1900-MHz hyperband, 50

Effective radiated power (ERP), 54
EIA/TIA-533, 2, 36
  defined, 2
  frequency modulation (FM), 4
Electronic serial number (ESN), 100, 274

Electronics Industry Association
    (EIA), 1–2, 36
Emissions masks, 56
  defined, 56
  illustrated, 57
Encryption
  on DCCH, 284–85
  signaling message, 282–84
  TMSI and, 284
Enhanced data for GSM evolution (EDGE)
  air interface, 336
Enhancements. *See* 136+; 136HS
Equipment testing, 307–23
  CTIA, 322–23
  interoperability, 314–16
  mobile, 309–23
  network, 308–9, 310
  performance, 311–13
  protocol conformance, 316–22
  reasons for, 307–8
Erlang B model, 23–24
Extended broadcast control channel
    (E-BCCH), 62, 63, 122
  Change (EC) field, 79, 180
  Cycle Length (ECL) field, 79
  frame definition, 80
  layer 2 frame types, 78–79
  messages, 94
  See also Broadcast control channel
    (BCCH)
Fast associated control channel
    (FACCH), 66, 88
  messages on forward DTC, 100–103
  messages on reverse DTC, 103–5
Fast broadcast control channel
    (F-BCCH), 62, 63, 122
  Change (FC) field, 79, 180
  frame definition, 80
  layer 2 frame types, 78–79
  messages, 92, 93
  See also Broadcast control channel
    (BCCH)
Federal Communications Commission
    (FCC), 2
Flash with Info Ack message, 101, 104
Flash with Info message, 101, 103–4
Forward DCCH (FDCCH), 84–85

Forward DCCH (FDCCH) (continued)
  Indication primitive, 74, 76
  mobile monitors, 84
  service access point (FDCCH SAP), 73
  shared channel feedback, 85
  See also DCCH
Frame number map (FRNO MAP), 78
Frame structure
  defined, 56–57
  TDMA, 58
Frequency division duplex (FDD), 2, 22
Frequency Division Multiple Access
  (FDMA), 1
Frequency shift keying (FSK)
  modulation, 2
Full rate, 21
Future Public Land Mobile
  Telecommunications System
  (FPLMTS), 331

General UDP transport service
  (GUTS), 14, 229–30
  defined, 229
  messages, 230
  thin client architecture, 229
Global challenge authentication,
  135, 276–78
  AUTH bit, 276
  defined, 276
  process illustration, 277
  See also Authentication
Global Operators' Forum (GOF), 329
Global System for Mobile
  Communications. See GSM
Global TDMA Forum (GTF), 326–28
  136+ Working Group, 327, 328
  activities, 326–27
  defined, 326
  HSD Working Group, 327, 328
  ITG Working Group, 327–28
Global WIN Forum (GWF), 328–29
GSM
  defined, 15
  Digital PCS relationship with, 15–16
  MAPs, 16

Hamming code, 66–68
Handoff

algorithms, 201
boundaries, 304–5
hyperband, 269
intersystem, 118
message, 102
message flows for, 200
mobile assisted (MAHO), 7, 15, 199
Hierarchical cell structures (HCS), 20
  ACA combined with, 28
  capacity increase, 26, 27
  defined, 24, 141
  in Digital PCS system, 25
  implementation, 24, 165–69
    example, 168
    rules, 166–67
  See also Reselection
Higher layer protocol identifier
  (HLPI), 32, 211–12
  defined, 211
  formatting, 211
  subfields, 211–12
High Speed Data (HSD) Working
  Group, 327, 328
High-tier mobiles, 120
Home location register (HLR), 4, 108,
  109, 113–14
  defined, 114
  stand-alone, 114
  for teleservice routing information, 208
  See also Visitor location register (VLR)
Hyperbands
  channelization in, 49, 50
  defined, 48
  duplexing in, 49, 50
  handoff, 269
  reselection, 267–69
  RF channels, 48
  spectrum allocation within, 49
Hyperframes
  defined, 59
  structure, 60, 64

Identity mobile station identity
  (IMSI), 136, 137–38
  defined, 138
  formatting and encoding, 138
IMT-2000, 331–33

defined, 331
spectrum allocations, 332
voice quality objectives, 332
Information elements
   defined, 90
   mandatory, 90
   Message Type (MT), 90
   optional, 90
   Protocol Discriminator (PD), 90
   R-Cause, 208–9
   R-DATA, 208
   SMS DELIVER/SMS SUBMIT
      messages, 219–20
Instantaneous capacity, 22
Intelligent roaming, 109–10
   800-MHz scanning, 263, 264
   defined, 255
   full-band scanning, 264–65
   methods of providing, 256
   mobile-directed, 256–57
   need for, 256
   network-directed, 256
   procedure definition, 263
   scanning process, 258
   triggered scanning, 265–67
Intelligent roaming database (IRDB), 227
   band order, 262
   defined, 257
   IR Control Data entries, 262
   parameters, 259–60
   reprogramming of, 267
   storage, 259
   updating, 257, 267
Interactive voice response (IVR), 116
Interim standard, 38
Interleaving, 68
International mobile station
      identity (IMSI), 16
International Organization for
      Standardization (ISO), 71
International Telecommunications Union
      (ITU), 13
Interoperability Test Group (ITG)
      Working Group, 327–28
Interoperability testing, 314–16
   defined, 314
   procedure, 314

summary, 315–16
See also Testing
Intersystem signaling, 269
Interworking function (IWF), 16, 107,
      112–13, 235, 241
   defined, 112
   functions, 113
   MSC integration, 113
IS-54, xiii–xiv, 6–9, 36
   AMPS channel comparison, 7
   calling number identification, 8
   defined, 6
   DTC, 6, 7–8
   enhancements, 8
   failure of, 8–9
   feature comparison, 10
   IS-136 vs., 39
IS-130, 43, 233
IS-135, 43
   data part, 238–40
   function model, 239
   services, 238
IS-136, 9–13, 39–41
   defined, 39
   Interoperability Test Specification, 314
   IS-54 vs., 39
   link layer, 39
   Revision 0, 9, 39
   Revision A, 9, 39, 40
   See also Standards; TIA/EIA-136
IS-137, 41–42
   defined, 41
   emissions masks, 56
   Revision 0, 41
   Revision A, 41–42
IS-138, 41–42
   defined, 41
   emissions masks, 56
   Revision 0, 41
   Revision A, 41–42
IS-641 vocoder, 15, 36, 42, 186–91
   adaptive prefilter, 186
   bit allocations, 189
   CELP, 186
   channel coding, 188–89, 190
   output, 188
   speech windowing, 188

IS-641 vocoder (continued)
    synthesis filter, 187

Layered protocol, 29–31
    data link layer, 30, 71–88
    network layer, 30, 89–105
    physical layer, 30, 47–69
Levinson-Durbin algorithm, 187
Long-term MACA (LTM), 136, 178
Long-term predication (LTP), 186
Low-tier mobiles, 120

MACA Report message, 99
Maintenance message, 102
Measurement Order Ack message, 104
Measurement Order message, 102
Media access control, 82–85
Message center, 12
Message encryption algorithm
    (MEA), 284–85
Message encryption key (MEK), 284–85
Messages
    BCCH, 92–95
    for call origination, 195
    for call termination, 197
    CMT, 218–23
    DCCH, 92–100
    DTC, 100–105
    expansion of, 92
    FACCH, 100–105
    GUTS, 230
    for handoff, 200
    information elements, 90
    mapping of, 91
    OATS, 223, 225, 227
    OPTS, 228
    RACH, 98–100
    SACCH, 100–105
    SPACH, 95–98
    See also Network layer
Message waiting indication (MWI), 13
Message Waiting message, 96
Mid-tier mobiles, 120
Mobile Ack message, 104
Mobile assisted channel allocation
    (MACA), 28, 94–95
    list, 180
    long-term (LTM), 136, 178, 300

message, 94–95, 136
    reporting, 28, 179
    short-term (STM), 136, 178, 300
    sleep mode and, 178–79
Mobile Assisted Channel Allocation
    message, 94–95, 178
Mobile assisted handoff
    (MAHO), 7, 15, 199
Mobile identification number
    (MIN), 2, 136–37, 274
Mobiles
    class I, 120
    class IV handhelds, 119
    dual-band, 270, 271
    dual-mode, 270, 271
    high-tier, 120
    low-tier, 120
    mid-tier, 120
    state diagram, 121
    types of, 119–20
Mobile states, 120–28
    Control Channel Scanning and
        Locking, 120–22
    DCCH Camping, 122–23
    defined, 120
    illustrated, 121
    Originated Point-to-Point Teleservice
        Proceeding, 127–28
    Origination Proceeding, 124–25
    Registration Proceeding, 123–24
    SSD Update Proceeding, 126–27
    Terminated Point-to-Point Teleservice
        Proceeding, 125–26
    Waiting for Order, 125
Mobile station (MS), 107, 109–10
    components, 109
    control unit, 109
    intelligent roaming, 109–10
    short message entity, 110
    transceiver unit, 109
    See also Mobile station procedures
Mobile station IDs (MSIDs), 76, 136–39
    defined, 136
    IMSI, 136, 137–38
    MIN, 136–37
    permanent (PMSIDs), 136
    TMSI, 136, 138–39

Mobile station power class, 54
Mobile station procedures, 128–36
   authentication, 135
   Control Channel Reselection, 130–31
   Control Channel Selection, 129–30
   DCCH Scanning and Locking, 128–29
   Originated Point-to-Point
       Teleservice, 132–33
   Origination, 132
   Registration, 133–34
   Registration Reject, 134–35
   Registration Success, 134
   Termination, 131–32
Mobile switching center
       (MSC), 4, 5, 6, 107
   AC integration, 114
   call-processing functions, 112
   functions of, 111
   IWF integration, 113
   OA&M functions, 112
   radio resource management, 112
Mobile testing, 309–23
Monitoring of radio link quality (MRLQ)
       counter, 87
MS_ACC_PWR, 296–99

Neighbor Cell message, 94, 142, 143
Neighbor channel measurements, 175–78
   interval increase conditions, 181
   intervals, 176–77
   number of channels and, 176
   performance of, 175–76
Neighbor list, 11
Network
   interfaces, 116–18
   overview, 107–9
   reference model, 108
Network layer, 89–105
   defined, 30, 89
   information elements, 90
   introduction to, 89–92
   message set, 89
   new features and, 30–31
   See also Layered protocol
Network parameter settings, 287–305
   access attempts, 299–302
   Access Burst Size, 289

access thresholds, 296–99
CELLTYPE, 290
DCCH location, 291–96
DCCH structure, 302–3
DELAY, 290
HL_FREQ, 290
importance of, 287–90
key, 288–90
Max Busy/Reserved, 289
Max Repetitions, 289
Max Retries, 289
Max Stop Counter, 289
MS_ACC_PWR, 289, 296–99
probability blocks, 288
RESEL_OFFSET, 290
RSS_ACC_MIN, 288, 296–99
SCANINTERVAL, 290
SS_SUFF, 290
sub-band priorities, 288
Network testing, 308–9, 310
Network types
   defined, 244
   identification, 244
   neighboring control channels marked
       with, 245
   priorities, 248
Nonpublic services, 243–53
   autonomous systems, 249–53
   network types, 243–45
   PSIDs, 245–49
   RSIDs, 245–49
North American Digital Cellular
       (NADC), 6
Number of additional words coming
       (NAWC), 318
Numeric assignment module
       (NAM), 16, 248

Open-loop pitch analysis, 187–88
Operations, administration, and
       maintenance (OA&M)
       functions, 111, 112
Optional information elements, 90–92
Originated Point-to-Point Teleservice
       procedure, 132–33
Originated Point-to-Point Teleservice
       Proceeding state, 127–28

Origination message, 99, 196
Origination procedure, 132
Origination Proceeding state, 124–25
OSI reference model, 71
 illustrated, 72
 layers, 72–73
 teleservices in, 205–6
Over-the-air activation (OAA), 12
 accomplishment, 223
 defined, 223
 network architecture, 224
 process, 223–27
Over-the-air activation function
   (OTAF), 224, 226, 227
Over-the-air activation teleservice
   (OATS), 115, 223–27
 defined, 223
 message flows, 225
 messages, 223, 227
 See also Teleservices
Over-the-air programming (OAP), 227
Over-the-air programming teleservice
   (OPTS), 115, 227–28
 defined, 227, 228
 messages, 228
 See also Teleservices

Packet control channel (PCCH), 333–34
Packet traffic channel (PTCH), 333–34
Page Response message, 99
Paging channel (PCH), 62, 122
 Displacement, 174, 175
 slots, number of, 303
 subchannels, 172–75
Paging frame classes (PFCs), 11, 172–74
 changing, 174
 defined, 172
 list of, 172
 PCH subchannels and, 173
PCS1900. *See* GSM
Performance testing, 311–13
 defined, 311
 parameters, 311–13
 reselection time, 312–13
 selection time, 311–12
 standby time, 313
 talk time, 313

See also Testing
Periodic Evaluation example, 165
 CEF5 calculation for, 166
 defined, 165
 See also Reselection
Physical layer, 47–69
 channel coding and interleaving, 66–69
 channelization and duplexing, 48–50
 defined, 30, 47
 frame and time-slot structures, 56–66
 modulation format, 51–54
 power output, 54–56
 See also Layered protocol
Physical Layer Control Ack message, 104
Physical Layer Control message, 102, 199
Power output, 54–56
 base station, 56
 requirements, 55
Power-up scan, 260, 261
Priority System example, 163–65
 CEF4 calculations for, 164
 defined, 163–64
 See also Reselection
Private operating frequencies
   (POFs), 250, 260
Private system IDs (PSIDs), 123, 245–49
 alpha tags, 249
 autonomous systems and, 253
 cell site support, 247
 defined, 245
 in forming closed user groups, 246
 partitioning of, 246
 priorities, 248
 ranges of, 247
 storing, 248–49
 See also Residential system IDs (RSIDs)
Probability blocks, 291–92
 800-MHz, 291–92
 defined, 292
Probability of locking, 22–23
Processed signal strength (PSS), 181, 182
Protocol conformance testing, 316–22
 defined, 316
 performance of, 317, 321
 plan coverage, 318–21
 requirement of, 317
 See also Testing

Public service profiles (PSPs), 251
Public switched telephone network
    (PSTN), 108, 235
RACH messages, 98–100
    Audit Confirmation, 98
    Authentication, 98
    Base Station Challenge Order, 99
    Capability Report, 99
    concatenation rules, 98
    MACA Report, 99
    Origination, 99, 196
    Page Response, 99
    R-DATA, 32, 96, 99, 102, 104, 128
    R-DATA ACCEPT, 96, 99–100,
        102, 104, 208
    R-DATA REJECT, 96, 100,
        102, 104, 208
    Registration, 100
    Serial Number, 100
    SPACH Confirmation, 100
    SSD Update Order Confirmation, 100
    Test Registration, 100
    Unique Challenge Order
        Confirmation, 100
Radio Link Failure example, 158–59
    CEF1 calculations, 159
    defined, 158
    See also Reselection
Radio Link Protocol (RLP1), 233
    data link connections, 240
    data link support, 235
    encoded frames, 237
    function model, 236
    functions, 237–38
Radio resource management, 112
Random access, 82
R-Cause, 208–9
R-DATA messages, 32, 96, 99, 102,
        104, 128
    ACCEPT message, 96, 99–100, 102,
        104, 208
    contents, 210
    information elements, 208
    length, 132–33
    REJECT message, 96, 100, 102,
        104, 208

Reauthentication Order Confirmation
    message, 104
Reauthentication Order message, 102
Received signal strength (RSS)
    measurement, 199
Registration message, 100
    Accept, 96–97
    Parameters, 93
    Reject, 97
Registration Proceeding state, 123–24
Registration-related procedures, 133–35
    Registration, 133–34
    Registration Reject, 134–35
    Registration Success, 134
    See also Mobile station procedures
Regulatory Configuration message, 94
Release message, 102, 104, 196
Reorder/Intercept message, 97
Reselection
    boundaries, 304–5
    control channel locking and, 143–44
    defined, 141
    examples, 157–65
        Cell Barred, 159–61
        Directed Retry, 162–63
        parameters for, 158
        Periodic Evaluation, 165
        Priority System, 163–65
        Radio Link Failure, 158–59
        Server Degradation, 161–62
    hyperband, 267–69
    neighbor cell information in, 142–43
    See also Hierarchical cell
        structures (HCS)
Reselection Criteria procedure, 143–44
Reselection trigger conditions
        (RTCs), 144–46
    Directed Retry, 144, 145, 146
    mapping into CEFs, 147
    Periodic Evaluation, 146
    Priority System, 145, 146
    Radio Link Failure, 145
    Server Degradation, 145
    Service Offering, 146, 153
Residential system IDs
        (RSIDs), 123, 245–49
    alpha tags, 249

Residential system IDs (continued)
   autonomous systems and, 253
   cell site support, 247
   defined, 245
   in forming closed user groups, 246
   partitioning of, 246
   priorities, 248
   storing, 248–49
   See also Private system IDs (PSIDs)
Reverse access control channel
      (RACH), 62, 84–85, 247
   layer 2 frames, 77, 78
   mobile access flowchart, 83
   subchannels on, 86
   See also RACH messages
Reverse DCCH (RDCCH)
   Indication primitive, 76
   Request primitive, 74
   service access point (RDCCH SAP), 73
   See also DCCH
RSS_ACC_MIN, 296–99

Serial Number message, 100
Server Degradation example, 161–62
   CEF2 calculations for, 161
   defined, 161
   See also Reselection
Service access points, 71
   DCCH layer 2, 73–76
   forward DCCH (FDCCH SAP), 73
   illustrated, 75
   reverse DCCH (RDCCH SAP), 73
Service data units (SDUs), 235
Service Menu message, 93–94
Service primitives, 71
   DCCH layer 2, 73–76
   service access points, 75
Service providers
   acceptable, 257
   favored, 257
   unacceptable, 258
Service Request message, 104
Service Response message, 102
Shared secret data (SSD), 95, 122
   attacks to find, 274
   defined, 275
   generation, 275

Update Order Confirmation
      message, 100, 104
Update Order message, 97, 102–3
Update procedure, 227
Update Proceeding state, 126–27
updating, 274, 275, 280–82
updating process, 280–81
Short Message Delivery–Point-to-point
      (SMDPP) message, 32, 207
   contents, 208, 209
   defined, 32
Short message entities (SMEs), 207
Short message service (SMS), 12, 96
Short-term MACA (STM), 136, 178, 179
Signaling message encryption, 8
Sleep mode, 171–82
   BCCH rereading and, 179–81
   defined, 171
   extending, 181–82
   MACA and, 178–79
   neighbor channel measurements
      and, 175–78
   processes impacting, 175–81
Slow associated control channel
      (SACCH), 88
   messages on forward DTC, 100–103
   messages on reverse DTC, 103–5
Smart antennas, 111
SMS broadcast control channel
      (S-BCCH), 62–63, 180, 214
   broadcast set for, 217
   changes to, 181
   subchannels, 180
   See also Broadcast control channel
      (BCCH)
SMS channel (SMSCH), 62
SMS paging, access response channel
      (SPACH), 62
   frame type, 82
   layer 2 frames, 80–82
   Request primitive, 76
   See also SPACH messages
SMS_TeleserviceIdentifier, 211–12
SOC/BSMC Identification message, 94
SPACH messages, 95–98
   Analog Voice Channel Designation, 95
   Audit Order, 95

Base Station Challenge Order
  Confirmation, 95
Confirmation, 100, 125
Digital Traffic Channel
  Designation, 95, 126
Directed Retry, 96, 196, 198
Message Waiting, 96
Notification, 123, 124, 125–26
R-DATA, 96
R-DATA ACCEPT, 96
R-DATA REJECT, 96
Registration Accept, 96–97
Registration Reject, 97
Reorder/Intercept, 97
SPACH Notification, 97
SSD Update Order, 97
Test Registration Response, 97
Unique Challenge Order, 97–98
Speech windowing, 188
Square-root raised cosine filter, 53
SS7 protocol, 116, 117
  global title translation (GTT), 222
  signaling transfer point, 222
  TIA/EIA-41 structure, 116, 117
  transaction capabilities application part
    (TCAP), 117
Standards, 35–45
  American National, 38
  development stages, 37
  Interim, 38
  list of, 35
  summary, 40
  See also specific standards
Standby time, 171
Stop Measurement Order message, 103
Sub-bands
  defined, 292
  priorities, 293–95
Subscriber identity module (SIM), 16
Superframes
  defined, 59
  full-rate DCCH slot allocations, 63
  PCH subchannels on, 173
  structure, 60, 64
Supervisory audio tone (SAT), 95
System identity (SID), 245, 257
System Identity message, 93

System operator code (SOC), 245, 257
TDMA, xiii, 7, 20–24
  Digital PCS increase with TDMA, 21
  example use of, 20
  frame structure, 58
  full-rate channel, 57
  timing offset, 59
  trunking efficiency, 22
Telecommunications Industry Association
  (TIA), 1, 36
Teleservices, 31–32, 205–30
  broadcast transport, 213–17
  carrier-specific teleservice, 206–7
  CMT, 115, 218–23
  defined (standardized), 206, 217–30
  defined, 12, 31–32, 205
  delivery of, 206, 207–10
  GUTS, 14, 229–30
  OATS, 115, 223–27
  OPTS, 115, 227–28
  in OSI model, 205–6
  reassembly, 115, 206, 213
  routing information, 208
  segmentation, 115, 206, 213
  support of, 15
  under TIA/EIA-136, 143
Teleservice segmentation and reassembly
  (TSAR), 115, 206
Teleservice server, 115–16
  functions, 115
  IVR link, 116
Temporary mobile station identity
  (TMSI), 136, 138–39
  addressing through, 139
  defined, 136
  encryption and, 284
Terminated Point-to-Point Teleservice
  Proceeding state, 125–26
Termination procedure, 131–32
Testing
  CTIA, 322–23
  equipment, 307–23
  interoperability, 314–16
  mobile, 309–23
  network, 308–9, 310
  performance, 311–13

Testing (continued)
    protocol conformance, 316–22
    reasons for, 307–8
Test Registration message, 100
Test Registration Response message, 97
Thin client architecture (TCA), 229
TIA/EIA-41, 2, 4, 36
    defined, 116
    facilities directive (FACDIR)
        message, 201
    facilities directive 2 (FACDIR2)
        message, 202
    handoff measurement request
        (HANDMREQ) message, 201
    MAP, 117
    mobile application part (MAP), 16
    mobile on channel (MSONCH)
        message, 201
    SS7-based protocol structure, 116, 117
TIA/EIA-136, xiii–xv, 9, 36, 43–45
    air interface, 326
    control and traffic channels, 36
    criteria, 147
    defined, 43
    layer definition, 73
    modulation format, 51–54
    parts, 44–45
    power output, 54–56
    teleservices under, 43
    See also IS-136
Time alignment, 58
Time division duplexing (TDD), 20
Time Division Multiple Access. *See* TDMA
Time-slot structures, 56–66
    DCCH, 59–63
    DTC, 63–66
TR-45.3, 36, 37
Trunking efficiency, 22

Unique challenge authentication, 278–80
    COUNT, 279–80
    defined, 278–79
    process illustration, 278
    RANDU, 279
    See also Authentication
Unique Challenge Order Confirmation
    message, 100, 105

Unique Challenge Order message,
    97–98, 103
U.S. Digital Cellular (USDC), 6
Universal Wireless Communications
    Consortium (UWCC), 314,
    325–26
    activities, 326
    board-level companies, 339–40
    defined, 325
    enhancements, 332
    general membership, 340–41
    member companies, 339–41
    objectives, 325–26
User datagram protocol (UDP), 229

Vector sum excited linear predictive
    (VSELP)
    coding, 7, 9, 10
    vocoders, 202
Visitor location register (VLR), 4, 5, 108,
    109, 113–14
    defined, 114
    user interface, 114
    See also Home location register (HLR)
Vocoders, 183–93
    assignment, 201–3
    bit-exact implementation, 193
    CELP, 186
    defined, 183
    fixed-point implementation, 193
    implementing, 193
    IS-641, 15, 36, 42, 186–91
    operation illustration, 184
    placement strategies, 193
    power of, 185
    speech synthesizer, 185
    VSELP, 202
Voice
    activity detection, 191–92
    coding/decoding, 183–93
    privacy, 8, 282–84
    services, 183–203
Voice privacy mask (VPMASK), 282–84
Voltage controlled oscillator (VCO), 270

Waiting for Order state, 125
Wireless application protocol
    (WAP), 229–30

Recent Titles in the Artech House Mobile Communications Series

John Walker, Series Editor

Advances in Mobile Information Systems, John Walker, editor

An Introduction to GSM, Siegmund M. Redl, Matthias K. Weber, Malcolm W. Oliphant

CDMA for Wireless Personal Communications, Ramjee Prasad

CDMA RF System Engineering, Samuel C. Yang

CDMA Systems Engineering Handbook, Jhong S. Lee and Leonard E. Miller

Cell Planning for Wireless Communications, Manuel F. Cátedra and Jesús Pérez-Arriaga

Cellular Communications: Worldwide Market Development, Garry A. Garrard

Cellular Mobile Systems Engineering, Saleh Faruque

The Complete Wireless Communications Professional: A Guide for Engineers and Managers, William Webb

GSM and Personal Communications Handbook, Siegmund M. Redl, Matthias K. Weber, Malcolm W. Oliphant

GSM Networks: Protocols, Terminology, and Implementation, Gunnar Heine

GSM System Engineering, Asha Mehrotra

Handbook of Land-Mobile Radio System Coverage, Garry C. Hess

Introduction to Radio Propagation for Fixed and Mobile Communications, John Doble

Introduction to Wireless Local Loop, William Webb

IS-136 TDMA Technology, Economics, and Services, Lawrence Harte, Adrian Smith, Charles A. Jacobs

Mobile Communications in the U.S. and Europe: Regulation, Technology, and Markets, Michael Paetsch

Mobile Data Communications Systems, Peter Wong, David Britland

Mobile Telecommunications: Standards, Regulation, and Applications, Rudi Bekkers and Jan Smits

Personal Wireless Communication With DECT and PWT, John Phillips, Gerard Mac Namee

Practical Wireless Data Modem Design, Jonathan Y.C. Cheah

RDS: The Radio Data System, Dietmar Kopitz, Bev Marks

RF and Microwave Circuit Design for Wireless Communications, Lawrence E. Larson, editor

Spread Spectrum CDMA Systems for Wireless Communications, Savo G. Glisic, Branka Vucetic

Understanding Cellular Radio, William Webb

Understanding Digital PCS: The TDMA Standard, Cameron Kelly Coursey

Understanding GPS: Principles and Applications, Elliott D. Kaplan, editor

Universal Wireless Personal Communications, Ramjee Prasad

Wideband CDMA for Third Generation Mobile Communications, Tero Ojanperä, Ramjee Prasad

Wireless Communications in Developing Countries: Cellular and Satellite Systems, Rachael E. Schwartz

For further information on these and other Artech House titles, including previously considered out-of-print books now available through our In-Print-Forever® (IPF®) program, contact:

Artech House
685 Canton Street
Norwood, MA 02062
Phone: 781-769-9750
Fax: 781-769-6334
e-mail: artech@artechhouse.com

Artech House
46 Gillingham Street
London SW1V 1AH UK
Phone: +44 (0)171-973-8077
Fax: +44 (0)171-630-0166
e-mail: artech-uk@artechhouse.com

Find us on the World Wide Web at:
www.artechhouse.com